W9-CRY-887

AMORPHOUS METALS

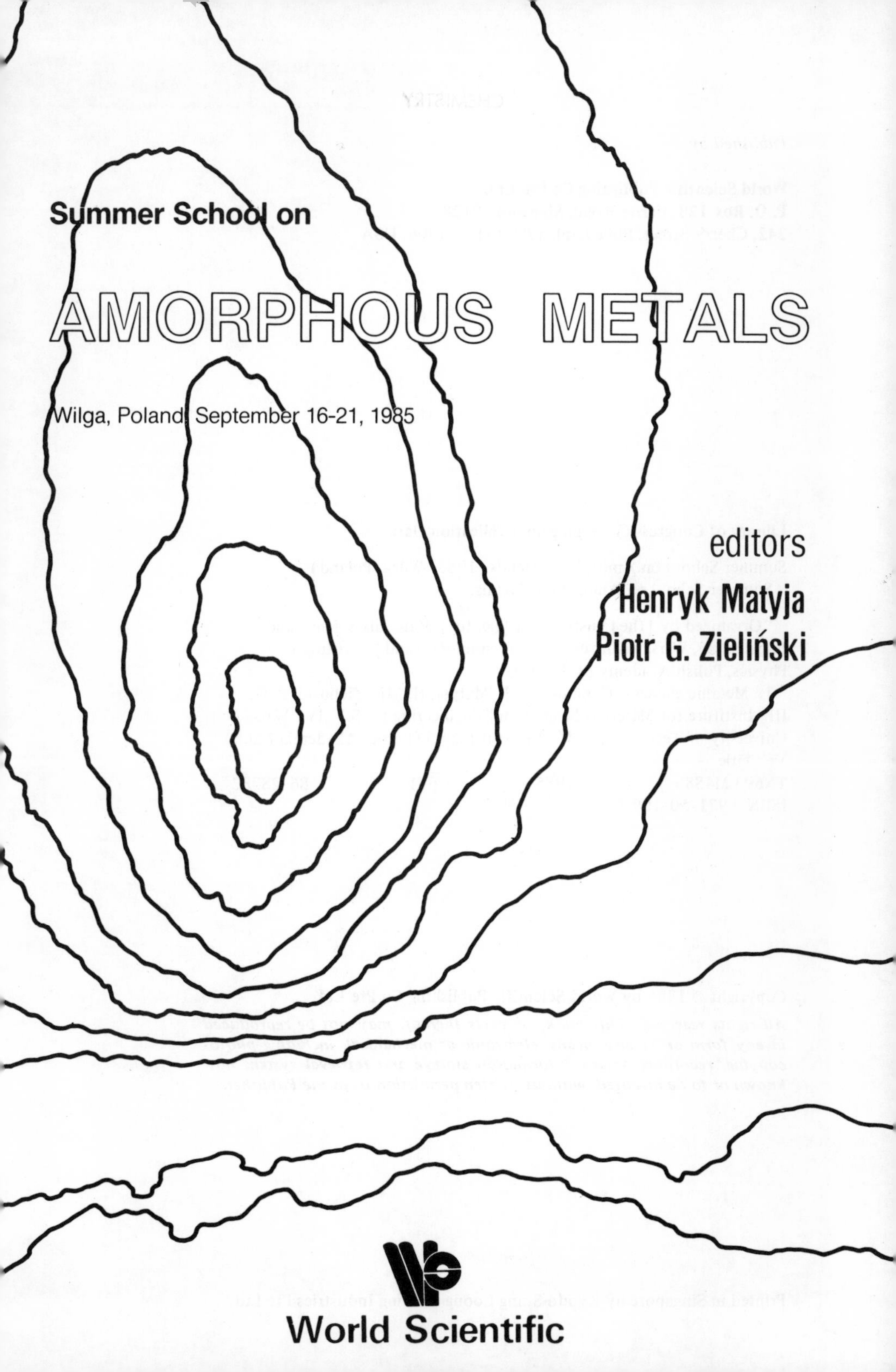

Summer School on

AMORPHOUS METALS

Wilga, Poland September 16-21, 1985

editors

Henryk Matyja

Piotr G. Zieliński

World Scientific

Published by

World Scientific Publishing Co Pte Ltd.
P. O. Box 128, Farrer Road, Singapore 9128
242, Cherry Street, Philadelphia PA 19106-1906, USA

Library of Congress Cataloging-in-Publication Data

Summer School on Amorphous Metals (1985 : Wilga, Poland)
Summer School on Amorphous Metals.

"Organized by [the] Institute of [i.e. for] Materials Science and Engineering, Warsaw University of Technology [and] Institute of Physics, Polish Academy of Sciences" —

1. Metallic glasses – Congresses. I. Matyja, H. II. Zielinski, P. G. III. Institute for Materials Science and Engineering (U.S.) IV. Warsaw University of Technology. V. Instytut Fizyki (Polska Akademia Nauk) VI. Title.

TN693.M4S86 1985 671 86-28282
ISBN 9971-50-109-0

Printed in Singapore by Kyodo-Shing Loong Printing Industries Pte Ltd.

Organizers

Institute of Materials Science and Engineering
Warsaw University of Technology
Institute of Physics, Polish Academy of Sciences

Organizing Committee

H. Matyja
M. W. Grabski
H. K. Lachowicz
H. Szymczak
S. Wojciechowski
P. G. Zieliński
W. Zych

Acknowledgements

The editors wish to express their sincere thanks to Mr. Boguslaw Dabrowski for the editorial work. The help of the staff of the Institute of Materials Science and Engineering and the Institute of Physics before and during the school is greatly appreciated.

CONTENTS

This book is dedicated to Dr P. G. Zieliński

Piotr G. Zieliński, one of the editors of this book, died suddenly on 27th of April 1986. He received a M.Sc. degree in 1972 and a Ph.D. in 1977 from the Warsaw Technical University. Following completion of M.Sc. studies he joined the Institute of Materials Science and Engineering of the Warsaw Technical University.

Dr Zieliński was the author of more than thirty papers concerning amorphous metals well known in the international scientific community. He presented at the IInd International Conference on Rapidly Quenched Metals at the Massachusetts Institute of Technology in 1975 a paper on the glass forming ability, being a part of his Ph.D. thesis. The years 1981-83 he spent at the Department of Materials Science and Engineering of the Cornell University investigating the plastic deformation of metallic glasses. He was greatly talented scientist and organizer. The participants of the Summer School on Amorphous Metals at Wilga will remember for long his great and versatile activity there.

H. Matyja

PREPARATION OF METALLIC GLASSES AND ITS INFLUENCE ON THE PROPERTIES

J. Bigot
Centre d'Etudes de Chimie Metallurgique,
C.N.R.S., 15, rue Georges Urbain,
95507 VITRY/Seine Cedex

1. INTRODUCTION

Since the discovery of P. Duwez in the sixtienth[1], the preparation and study of amorphous metallic materials have been the subject of numerous works. New methods of preparation have been elaborated, most of the amorphous compositions have been known and their properties have been identified.

Now it looks rather evident that the properties of these materials are greatly influenced by the method and the conditions of production. Each particular process produces a material quality which bears the traces of the preparation conditions not only in its morphology but also in its structure and properties.

The preparation methods of well studied parameters are limited. Some of them have a potential industrial interest, namely, the planar flow casting, the sputtering process or laser melting.

In the present paper we will briefly recall the main methods for preparation of amorphous metallic materials. We will consider more extensively the continuous processes of melt spinning and of planar flow casting. Finally, we will analyse the influence of the preparation parameters on the resulting properties of the material.

2. MAIN PROCESSES OF PREPARATION OF METALLIC AMORPHOUS MATERIALS

2.1. Liquid Quenching Processes

If a liquid of suitable composition is cooled fast enough, the crystallization may be avoided. The rate of quenching is important, generally exceeding 10^6 K/s, but every composition has its own critical cooling rate[2] (e.g. 10^6 K/s for $Fe_{83}B_{17}$, 10^3 K/s for $Pd_{82}Si_{18}$)
The first process is the gun technique of Duwez where small droplets of liquid are accelerated and projected onto a cold surface[3] (Fig. 1). The next is the splat cooling process[4] where a drop of molten liquid is smashed between two rapidly moving copper discs (Fig. 2). Numerous compositions of amorphous alloys were prepared using this technique[5] (Table 1).

Only small samples can be produced in these two processes. Soon afterwards, people tried to prepare larger samples by continuous casting. Different methods have been put forward (Fig. 2).

One can project a jet of liquid metal onto the outside surface of a rotating cylinder[6], which is called the melt spinning process. As a variant we can also run the jet inside the surface of the rotating cylinder. The centrifugal force ensures a better contact between the ribbon and the substrate[7]. Chen and Miller projected the melt between a pair of rapidly rotating rollers: this gives a symmetrical quenching rate and is a very effective process[8]. In the melt extraction process the surface of the wheel catches and ejects the liquid from a nozzle[9] whereas in the melt drag process a ribbon is entrained out of the surface of the melt[10].

In the early eighties Narasimhan used the planar flow casting technique to produce wide ribbons[11].

In the same class of liquid quenching techniques there is also the laser melting in which ideal thermal transfer between melt and substrate gives high cooling rate[12]. For completeness, the plasma spray process should be mentioned[13].

2.2. Vacuum Evaporation Method

This method, sometimes called vapour quenching technique, gives thin films of a few tenth of micrometers which are studied by X-ray diffration, resistivity or electron microscopy[14].

These materials are produced under high vacuum with the substrate cooled by liquid nitrogen (Fig. 3). The vapours different elements are produced separately by electron guns. Film thickness and composition are monitored by two regulation systems. The main impurity of the vacuum is water vapour which is kept near 10^{-9} Tr.

Numerous amorphous alloys are obtained by this way, namely, CrAg, CoAg, CuMg, CuAu, FeAu, FeSn. M.R.Bennet has prepared thin films of transition elements: Fe, Ni, Co. Only cobalt was obtained in the amorphous state. It seems that small quantities of impurities are necessary to prepare a nominally pure amorphous phase of these metals. This may explain the differences between the crystallization temperatures which, according to different authors, are found to be 5 K, 40 K and 90 K, respectively[15)].

2.3. Cathode Sputtering Process

By this technique a gas at a pressure of 10^{-1} to 10^{-9} Tr is ionized under a potential difference of 1 to 5 kV applied between two electrodes. The cathode which is also the target attracts the positive ions, which induces by bombardment, the sputtering of the atoms. The sample is produced by condensation of the sputtered atoms onto the anode. The discharge gas is generally an inert gas like argon which,in addition to its inertness, has a mass comparable to the one of the target atoms and hence ensures an efficient sputtering (Fig. 4).

Various set-ups are used,among them we may mention: the diode system with or without the use of radiofrequency, the triode system where ionization is obtained by electron bombardment [16, 17)].

A large number of amorphous alloys are prepared by these techniques. The amorphous alloys obtained by quenching from the liquid can also be prepared by sputtering and the range of compositions which can be obtained in the amorphous state is even larger than in the case of quenching from melt. For example:

$Fe_{1-x}B_x$: sputtering $0,09 < x < 0,49$
liquid quenching $0,14 < x < 0,25$ [18)]

$Co_{1-x}Zr_x$: sputtering $0,07 < x < 0,2$
liquid quenching $x \sim 0,1$ [19)].

This difference may be probably due to the hydrogen content in the sample. Some analyses give more than 1 at.% of this gas in the sample.

The preparation parameters are numerous: the temperatures of target and sample, the gas purity, geometrical arrangement in the apparatus, and argon pressure. Figure 5 shows these effects[20]. The magnetization is not changed by variation of the argon pressure while the coercitive force is very sensitive to these variations. Our knowledge of the influence of the parameters of this process is only at its beginning stage.

2.4. Production of Amorphous Alloys by Ion Implantation

It has been known for several years that ion implantation may be a suitable technique to produce amorphous alloys[21]. Ions of 50 to 500 keV penetrate into the metal and lose their energy in a series of either elastic or inelastic collisions until stopping. The final distribution of ions is best visualised as a Gaussian, R_p (average projected range) being the maximum value. For example, phosphorous implanted in Ni with an energy of 50 keV produces R_p of 300 Å. Current measured at the target gives an appropriate conversion of the number of ions stricking the target. Mass discrimination results in ultra-pure ion beams and combined with high vacuum in the chamber ensures the contamination free samples.

The practical methods for characterization of the amorphous phase are confined to electrical measurements and electron diffraction. Transition metals like Fe, Co, Ni form amorphous phases under P^+B^+ implantations.

Ion implantation gives a surface amorphization with amorphous layer thickness varying from 100 to 1500 Å. It seems that the critical concentration for amorphous alloy formation is the same for low temperature implantation as for liquid quenched alloy: 20 to 25 % for PdSi, NiP [22,23].

2.5. Formation of Amorphous Phase by Ions Mixing

In the implantation process the ions are introduced into the crystalline target as long as the amorphous concentration is obtained. In the ion beam mixing, a multi-layer film is first deposited onto an inert substrate with the average composition of the amorphous phase. This film is then irradiated with ions which are generally Xe^+ accelerated at 300 keV to the dose of $5x10^{14}$ Xe/cm^2 to produce amorphization[24]. Thickness of the samples was approximately the average projected range R_p.

Amorphous alloys like $Au_{71}Si_{29}$, $Pd_{80}Si_{20}$, $Pt_{80}Si_{20}$, $Au_{17}Co_{73}$, AuGe were obtained in this way.

2.6. Electrochemical Methods

The number of glassy metals obtained by electrolysis appears to be very limited. We may quote NiP, CoP, FeP. These alloys are extremely brittle and probably with a high hydrogen content[25].

2.7. Preparation by Solid State Reaction

Amorphous alloys may be obtained by diffusion at low temperatures[26]. Samples are prepared either by alternately evaporated crystalline films or by sandwich rolling of thin foils. Amorphization is obtained after annealing low temperature. This is evidenced by X-ray diffraction and electron microscopy[27]. Amorphous powders can also be directly obtained by a method now called the mechanical alloying, performed by high energy ball milling[28]. Amorphous state is obtained when one of the elements presents an anomalous fast diffusion in the other and high negative heat of mixing. Those conditions are met in systems like ZrNi, HfNi, CoZr[29].

3. PREPARATION OF AMORPHOUS RIBBONS BY THE MELT SPINNING PROCESS

The melt spinning is the most common process used in laboratories probably because the apparatus is very easy to build. In this technique, a jet of liquid metal impinges onto the outer surface of a rotating wheel to form a ribbon. The diameter of the orifice, at the end of the nozzle is of the order of a millimeter or less, the distance to the wheel is of a few millimeters and the tangential linear speed of the substrate is of an order of 10 to 50 m/s. Parameters of the process that influence the shape of the ribbon are well known.

The thickness of the ribbon increases with the ejection pressure and decreases with the linear speed of the substrate (Fig. 6)[30]. The width of the ribbon depends on the volumetric flow rate, i.e. on the section of the nozzle and on pressure[31]. Tilting of the nozzle increases the thickness[32]. Surrounding atmosphere and surface quality of the wheel seem influence the morphology of the ribbon and the quenching rate[33].

The observation with high speed camera shows that the ribbon emerges from a puddle of a few millimeters width and height (Fig. 7). The process of formation of the ribbon is very complex[34]. A liquid layer is dragged out of the puddle and next solidifies: momentum transport acts alone. Thermal transport produces a solid layer in the puddle in contact with the cold substrate. The thickness of this layer increases along the length of the puddle and the ribbon is dragged out on the moving substrate. These two processes probably take place and numerous works have been done on the subject. It appears that the different models give relations in which the parameters cannot be experimentaly differentiated[35).]

Hillman has shown that the dimensions of the ribbon are essentially governed by the length and width of the puddle. The length of the puddle is smaller when the wheel speed increases while the thickness of the ribbon decreases[32).]

Vincent[35)] suggests the following empirical relationship for the thickness of the ribbon:

$$t \propto Q^{A} V^{-B} \Theta^{m}$$

where:

t - thicknes, Q - volumetric flow rate, V - velocity of substrate, $\Theta = l/V$, l-puddle length.

The average quenching rate, R, may be related to the thickness of the ribbon by:

$$R \propto t^{-1} \quad \text{or} \quad R \propto t^{-2}$$

according to a Newtonian cooling or an ideal cooling[36).]

However, when the speed of the wheel increases, the heat transfer coefficient also increases[37)] and the thickness of the ribbon is not an only factor which influences the quenching rate[38).]

In the melt spinning process the width of the ribbons obtained cannot exceed 5 or 6 millimeters, moreover, this width is difficult to control owing to the instability of the jet and the puddle. The section of the ribbon is generally not rectangular[39).] The quality of the edges is influenced by the surrounding atmosphere[33]. However, this process is very common, it gives amorphous material of good quality and one can change the quenching rate simply by changing the speed of the wheel.

4. LIQUID QUENCHING BY THE PLANAR FLOW PROCESS

Experimental conditions of this process have been described in the US Patent of Allied Chemical[11]. Being an industrial process, the information are thereby rather rare.

The principle of this technique is a modification of melt spinning in which the nozzle orifice is a slot which gives a liquid strip (corresponding to the width of the ribbon) from one centimeter up to fifteen centimeters. The nozzle is very close to the surface of the wheel, some tenth of a millimeter. There may be a slight difference between the levels of the upper and the lower lips. This can be obtained either by machining or by tilting. This arrangement allows the puddle to be constrained and stable.

S.C. Huang[40] has identified the main parameters acting on the thickness and on the morphology of the ribbon. Figure 8 shows the variations of the thickness against the ejection pressure for different nozzle slot breadth and with a constant wheel-to-nozzle gap. For low pressures, the thickness of the ribbon varies as the square root of the pressure. This is in agreement with the Bernouilli's law of flow. An interesting fact is that when the pressure increases, we reach a state of saturation, the presence of the wheel producing a modification of the flow. The free surface of the so obtained ribbon is better and the wheel side shows a wetting pattern which is different, with small air pocket defects only. The saturation effect appears sooner for narrow slot.

Thickness variation can be obtained by changing the wheel-to-crucible gap which changes also the slope of the curve in Fig. 8. Studies with high speed camera show that the length of the puddle increases when the pressure, the distance between the wheel and the crucible and the tilting increase[41]. Like in the melt spinning, an increase in the speed of the wheel results in the decrease of thickness and modifies the distribution of air pockets on the wheel side of the ribbon. However, the quality of the edges is not as good.

There are several publications dealing with the quenching rate of the ribbons. Vogt's results, obtained by computer calculations, show the variation of the quenching rate across the ribbon thickness for different transfer coefficients[42]. The overall variation is of 8 % for a ribbon of 25 μm and of 17 % for a ribbon of 50 μm. These calculations show also that the quenching rate is greater at the edges of the ribbon. This variation of the cooling rate is probably at the origin of the anisotropy of magnetic and mechanical properties of the ribbons.

An important parameter in the planar flow casting is the sticking length of the ribbon on the wheel after it formation under the puddle. This length can be varied from a few centimeters up to one turn of the wheel and increases more quickly than in the melt spinning. We found different reasons for this, the main one being the surface temperature of the wheel. This augmentation may a rise from the time of the run, the thermal conductivity of the wheel and of the wheel diameter. The sticking length increases more rapidly on a small diameter wheel because the time between two passes under the nozzle is shorter so that the surface temperature rises more quickly. The chemical composition of the ribbon alloy and of the wheel both act on the sticking length.

We know much about the parameters in planar flow casting, however, much more work is still necessary to improve the morphology of the ribbons and the quality of their surface.

5. INFLUENCE OF THE CONDITIONS OF PREPARATION ON THE PROPERTIES OF LIQUID QUENCHED RIBBONS

Among the first studies on this subject we find the work of Tagaki and Toi[43] on the preparation of $Fe_{40}Ni_{40}B_{20}$ alloys by the melt spinning technique. They showed that for this composition the Curie temperature T_c, the crystallization temperature T_x, magnetization M_s and hardness were not influenced by the speed of the wheel and the ejection temperature of the liquid metal. On the other hand, the coercitive force and permeability were found to be very sensitive to these parameters. Further studies of this kind were undertaken again afterwards for materials of various compositions.

Luborsky et al.[44] worked on FeNiB, FeB, FeSiB alloys covering a large domain of wheel speeds (Fig. 9). They showed that there is a critical speed at which completely amorphous samples can be obtained: they are also ductile and of low coercivity. For lower speeds, some crystallization takes place which increases H_c and induces brittleness. For higher speeds H_c is increased, probably owing to the fluctuation of residual strains. The effect of annealing on the properties depends also on the thickness of the ribbon and hence on the quenching rate. Magnetic alloys are generally very sensitive to the quenching rate variations[45,46).

The preparation conditions have various effects on the amorphous alloy crystallization. The published results show that temperature and enthalpy of crystallization of metal-metalloid alloys are not affected. However, glass transition of FeNiPb alloys increases when the quenching rate increases[47)]. For metal-metal alloys the effects are variable. In the NiZr system and for $Ni_{36,5}Zr_{63,5}$ T_x depends on the ejection temperature of liquid metal but not on the speed of the wheel. This is ascribed to variation of the chemical short range order when the liquid temperature increases. T_x of $Ni_{67}Zr_{33}$ is found independent of these parameters[48)]. For $Cu_{60}Zr_{40}$ alloys prepared by planar flow casting, an increase of the temperature of the liquid alloy or of the ejection pressure induce higher T_x[49)].

In the overall process of crystallization of amorphous metallic alloys, nucleation of the crystalline phase seems to be very sensitive to the quenching parameters, the rate of growth being not so modified[50)]. According to Greer[51)] in kinetic studies the number of nuclei in $Fe_{80}B_{20}$ is proportional to the inverse square of the quenching rate. Results of Köster[50)] in direct observation of FeNiB alloys showed a heterogeneous nucleation. The number of nuclei is found to decrease when the quenching rate increases and it is more important on the contact side than near the free side of the ribbon (Fig. 11). The nature of the wheel itself seems to act on crystallization at the contact side of the ribbon.

It is found for $Cu_{60}Zr_{40}$ alloys produced under different conditions that incubation time for crystallization determined by isothermal DSC is longer for alloys with faster quench. The number of quench nuclei decreases when the quenching rate increases and the crystallization peak in DSC is narrower[52).

Structural relaxation of amorphous metallic alloys is greatly influenced by the preparation conditions. The methods of study are numerous, they use the properties that are sensitive to these variations, namely: Curie temperature[53), magnetic properties, stress relaxation[54), brittleness, etc. Most of the works consider the variation of the properties after annealing at temperatures lower than T_g. Results show that an increase of quenching rate gives after the quench a material slightly relaxed corresponding to an important free volume and then very sensitive to further annealing. For such a material the relaxation is correspondingly greater and takes place at lower temperature[55).

The structural state of metallic glasses produced by planar flow casting is strongly dependent on the sticking length of the ribbon on the wheel[56). Hence FeNiB alloys are more ductile as the sticking length is longer. We observe the same trend for FeSiB alloys. On further annealing, decreasing the free volume induces brittleness, the secondary cooling will be therefore more important when the ribbon sticks on the wheel than when it cools only in the surrounding atmosphere. This fact is verified for ribbons that are produced under reduced pressure and a very low resultant ductility is observed[57).

The DSC technique which allows a measurement of ΔH_{relax} is a very good tool for that kind of studies. They show that even with a constant sticking length

ΔH_{relax} increases during the run which seems to be correlated to the overheating of the wheel. In this case also this variation is dependent on the nature of the wheel. There is a maximum temperature of the surface of the wheel where ΔH_{relax} is maximum (Fig.12). At more elevated temperatures some crystallization traces appear on the side of the ribbon contacting with the wheel[57)]

It seems that the effects that are usually ascribed to the mean rate of quenching could as well be ascribed to a variation in the secondary quench or to the sticking length rather than to the cooling rate under the puddle.

6. COMPARISON BETWEEN LIQUID QUENCHED AND SPUTTERED AMORPHOUS ALLOYS

Experimental results on samples of the same chemical composition but obtained in a different way are only a few. Generally, some of the properties like crystallization temperature, Curie temperature or electrical resistivity are very similar. However, careful studies show great differences between samples.

Harmelin and al. studied comparatively CuZr alloys prepared by liquid quenching and by sputtering[58)]. DSC curves obtained at 80 K/min were markedly different for the two samples(Fig. 13). Irreversible relaxation is more important for the sputtered samples when the annealing temperature is close to T_g, while the intensity of the endothermic effect in the glass transition range is more important for liquid quenched alloys.

Small angle scattering for the two samples are also different. In the as received sample the flatness of the Laue scattered intensity is an evidence for good homogeneity of the quenched sample. For the sputtered sample a weak broadening at very small angles indicates the presence of some large defects. Under annealing there is no such variation for the quenched sample. Whereas there is a maximum in the scattering which shifts to small angle values when the annealing temperature increases. The authors were not able to determine the exact nature of the scattering entities but they think that it may be in relation with the hydrogen content in the sputtered sample.

BIBLIOGRAPHY

1) Klement, J.W., Willen, R.H., Duwez, P., Nature 187, 869 (1960).
2) Davies, D.H.A., Proc. 3rd Int.Conf. Rapidly Quenched Metals, Brighton 1, 1 (1978).
3) Duwez, P., Willen, R.H., Klement, J.W., J.Appl.Phys. 31, 1136 (1960).
4) Dixmier, J., Guinier, A., Rev. Metall. 64, 53 (1967).
5) Takayama, S., J.Mater. Sci. 11, 164 (1976).
6) Pond, R.B., US Patent 2 825 108, March 4 (1958).
7) Pond, R.B., Maddin, R., Trans.Met.Soc. AIME 245, 2475 (1969).
8) Chen, H.S., Miller, C.E., Rev. Sci. Instrum. 41, 1237 (1970).
9) Karesh, S., US Patent 3 845 805 (1974).
10) Hubert, J.C., Mollard, F., Lux, B., Metallkde 64, 835 (1973).
11) Narasimhan, M., US Patent 4 142 571 (1979).
12) Miura, H., Isa, S., Omuro, K., Proc. 4th Int. Conf. Rapidly Quenched Metals, Sendai, 1, 43 (1981).
13) Gaffet, E., Deluze, G., Martin, G., Pelletier, J.M., Pergue, To be published in Mat.Sci.Eng.
14) Marchal, G., Maugin, Ph., Janot, C., Mag. 32, 1007 (1975).
15) Bennet, M.R., Wright, J.G., Phys. Status Solidi, A13, 135 (1972).
16) Bessot, J., Surfaces, n°95, p. 3 (1973).
17) Bunshah, R., Deposition technologies for films and coating, Noyes Publ. Park Ridge, N.J. U.S.A. (1982).
18) Shimada, Y., Kojima, H., J.Appl.Phys. 50, 3, 1541 (1979).
19) Shimada, Y., Kojima, H., J.Appl.Phys. 53, 4, 3156 (1982).

20) Heiman, N.,Hempstead, R.D., Asama, N.K., J.Appl. Phys. 49 11, 5663 (1978).

21) Ali, A., Grant, W.A., Grundy, J.P., Phil.Mag. B37, 353 (1978).

22) Cohen, C., Drico, A.V., Bernas, H., Chaumont, J., Krolas, K., Thome, L., Phys. Rev. Lett. 48 17, 1183 (1982).

23) Thome, L., Bengagoub, A., Audouard, A., Chaumond,J., J.Phys. F: Met.Phys. 15, 1229 (1985).

24) Mayer, J.W., Staub, B.Y., Lau, S.S., Hung, L.S., Nucl. Instrum. Methods 182, 1-13 (1981).

25) Flechon, J., Machizaud, F., Kuhnast, F.A., Baida, A.D., Rashid, A., Bul.Soc.Chim. France 7, 1257 (1982)

26) Schwarz, R.B., and Johnson, W.L., Phys.Rev.Lett. 51, 415 (1983).

27) Schröder, H., Samwer, K., Köster, U., Phys.Rev.Lett. 54 3, 197 (1985).

28) Koch, C.C., Cavin, D.B., Mc Kanney, C.G., and Scarbrough, J.D., Appl.Phys.Lett. 43, 1017 (1983).

29) Schultz, L., Proc. 5th Int. Conf. Rapidly Quenched Metals, Würzburg 2, 1585 (1984).

30) Takayama, S., and Oi, T., J.Appl.Phys. 50, 4962 (1979).

31) Liebermann, H.H., et Graham, C.D., IEEE Trans.Mag. MAG 12, 6, 921 (1976).

32) Hillman, H., Hilsinger, H.R., Proc. 3rd Int. Conf. Rapidly Quenched Metals, Brighton 1, 22 (1978).

33) Liebermann, H.H., Proc. 3rd Int. Conf. Rapidly Quenched Metals, Brighton 1, 34 (1978).

34) Kavesh, S., Metallic glasses. American Soc. Metals, Metals Park, Ohio (1978).

35) Vincent, J.H., Herbertson, J.G., Davies, H.A., Proc 4th Int.Conf. Rapidly Quenched Metals, Sendai 1, 77 (1981).

36) Ruhl, R.C., Mater.Sci.Eng. 1, 313 (1967).
37) Huang, S.C. and Fiedler, H.C., Mater.Sci.Eng. 51, 39 (1981).
38) Gillen, A.G., and Cantor, B., Acta Metall. 33 10, 1813 (1985).
39) Liebermann, H.H., Mater.Sci.Eng. 43, 203 (1980).
40) Huang, S.C., Proc. 4th Int. Conf. Rapidly Quenched Metals, Sendai 1, 65 (1981).
41) Scwartz, F., Thesis, Paris VI (1986).
42) Vogt, E., Frommeyer, G., Proc. 5th Int. Conf. Rapidly Metals, Würzburg 1, 63 (1984).
43) Takayama, S., and Oi, T., J.Appl.Phys. 50 3, 1595 (1979).
44) Luborsky, F.E., Liebermann, H.H., Walter, J.L., Metall. Glasses Sci. Techno. Budapest, 203 (1980).
45) Kulik, T., Lisowski, B., Zielinski, P.G., Matyja, H., Proc. 5th Int. Conf. Rapidly Quenched Metals, Würzburg 1, 1203 (1985).
46) Grossingerea, R., Sassik, H., Schotzko, Ch.,Lovas, A., Proc. 5th Int. Conf. Rapidly Quenched Metals, Würzburg 1, 1255 (1985).
47) Baricco, M., Battezzati, L., Marino, F., and Riontino, G., Proc. 5th Int. Conf. Rapidly Quenched Metals, Würzburg 1, 239 (1985).
48) Altounian, Z., and Strom-Olsen, J.O., Proc. 5th Int. Conf. Rapidly Quenched Metals, Würzburg 1, 447 (1985).
49) Wu, O.C., Harmelin, M., Bigot, J., Martin, G., J. Mat.Sci. 21 (1986).
50) Köster, U., Scripta Met. 17, 867 (1983).
51) Greer, A.L., Acta Met. 30, 171 (1982).
52) Harmelin, M., Calvayrac, Y., Quivy, A., Chevalier, J.P., Bigot, J., Scripta Met. 16, 703 (1982).
53) Greer, A.L., J. of Mat. Sc. 17, 1117 (1982).

54) Luborsky, F.E., and Walter, J.L., Mat. Science and Eng. 35, 255 (1978).

55) Battezzati, L., Riontino, G., Baricco, M., Lucci, A., and Marino, F., Proc. 5th Int. Conf. Rapidly Quenched Metals, Würzburg 1, 877 (1985).

56) Hilsinger, H.R., Hock, S., Metall. Glasses Sci. Technol. Budapest 1, 71 (1980).

57) Bigot, J., Harmelin, M., LAM VI, Garmisch Partenkirchen (1986) to be published.

58) Harmelin, M., Naudon, A., Frigerio, J.M., Rivory, J., Proc. 5th Int. Conf. Rapidly Quenched Metals, Würzburg 1, 659 (1985).

Table 1 - Composition of some amorphous metallic phases[5)]

System	Composition range x(at.%)	Method
$Au_{100-x}Si_x$	18.6-30	SQ
$Au_{75}Pb_{25}$	-	SQ
$Au_{100-x}Sn_x$	29-31	SQ
$Au_{73}Ge_{27}$	-	SQ
$Au_xGe_ySi_z$	x = 74-79 y = 12.4-13.6 z = 8.4-9.4	SQ
$Ag_{100-x}Si_x$	17-30	SQ
Ag_xCu_y	x = 35-65 y = 35-50	EV
$Ag_{100-x}Mn_x$	4-13	EV
$Al_{82,7}Cu_{17,3}$	-	SQ
$Al_{100-x}Ge_x$	30-80	EV
Ge, Te, B	Pure	EV, SQ
As, Bi, C, Ga, Se, Sb, Si	Pure	EV
Fe, Al, Cr, Pd	Pure	EV
Ni	Pure	EV, SQ
W, Mo, Ta, Nb, V	Pure	EV
Hf, Zr, Re	Pure	EV
$Cd_{25}Ge_{25}As_{50}$	-	EV
$Co_{100-x}Au_x$	25-65	EV
$Co_{100-x}P_x$	18-25	ED
$Co_{73}P_{16}B_{12}$	-	P
$Cu_{100-x}Ti_x$	30-35	SQ
Cu_xZr_y	x = 60,57 y = 40,43	SQ
Cu-Bi		EV

$Fe_{84}C_{16}$	-	SQ
$Fe_xP_yC_z$	x = 75-81 y = 10-16 z = 5-9	SQ, P
$Fe_{80}P_{13}B_7$	-	P
$Fe_{76}B_{13}C_7$	-	P
$Fe_{40}Ni_{40}P_{14}B_6$	-	SQ
$(Fe_{100-x}Mn_x)_{75}P_{15}C_{10}$	0-10	SQ
$Fe_{75}P_{16}Si_6Al_3$	-	SQ
$Fe_{75}P_{15}C_3Al_4$	-	SQ
$Fe_{77}P_{17}C_4Al_4$	-	SQ
$Fe_{74}P_{16}C_5Al_3Si_2$	-	SQ
$Fe_{76}P_{16}C_4Al_2Si_2$	-	SQ
$Fe_{72}P_{16}C_5Si_2Al_5$	-	SQ
$Fe_{80-x}Cr_xP_{13}C_7$	0-10	SQ
$Fe_{80-x}Ni_xP_{13}C_7$	0-40	SQ
$(Fe_{100-x}Ni_x)_{75}P_{15}C_{10}$	0-50	SQ
$Fe_{38\ 5}Ni_{38\ 5}P_{20-x}B_xAl_3$	2-10	SQ
$(Fe_{50}Ni_{50})_{80-x}P_{14}B_6A_x$ (A = Si or Al)	1-3	SQ
$(Fe_{100-x}Ni_x)_{77}P_{14}B_6Al_3$	0-10	SQ
$(Fe_{50}Ni_{50})_{81-x}P_{16}B_xAl_3$	4-7	SQ
$(Fe_{50}Ni_{50})_{91-x}P_xB_6Al_3$	14-17	SQ
$Fe_{80-x}P_{16}C_xB_1Al_3$	3-6	SQ
$Fe_{76\ 6}P_{14\ 2}C_{1\ 2}B_{4\ 8}Al_3$	2 -	SQ
$Fe_{70-x}Cr_{10}Ni_xP_{13}C_7$	5-20	SQ
$(Fe_{75}P_{15}C_6Al_4)_{100-x}(Fe_{77}P_{14}B_6Al_3)_x$	0-10	SQ
$Gd_{100-x}Co_x$	39-96	SQ
$Gd_{100-x}Fe_x$	15-94	SQ, EV
$Ge_{100-x}Bi_x$	0-25	EV
$La_{100-x}Au_x$	0-40	SQ
La-Cu	?	SQ
La-Ni	?	SQ

$Mg_{65}Cu_{35}$	–	EV
$Mg_{60}Sb_{40}$	–	EV
Mg-Bi	?	EV
$Mn_{100-x}Si_x$	23-28	SQ
$Mn_{75}P_{15}C_{10}$	–	SQ
$Nb_{100-x}Ni_x$	33-78	SQ
$Nb_{48}Ni_{39}Al_{13}$	–	SQ
$Ni_{100-x}P_x$	8.6-26.2	CD,ED,EV
$Ni_{100-x}Ta_x$	35-45	SQ
$Ni_{80}S_{20}$	–	ED
Ni-B	?	CD
$Ni_{75}P_{15}B_{10}$	–	P
$(Ni_{41}Pd_{41}B_{18})_{100-x}Cr_x$	0-4	SQ
$Ni_{90-x}P_xB_7Al_3$	16-18	SQ
$Ni_{49}Fe_{29}P_{14}B_6Al_2$	–	SQ
$Ni_{72}P_{14}B_6Si_3Al_5$	–	SQ
$Pb_{75}Au_{25}$	–	SQ
$Pb_{52}Sb_{48}$	–	SQ
$Pd_{100-x}Ge_x$	18-20	SQ
$Pd_{100-x}Si_x$	15-23	P, SQ
$Pd_{80-x}Cr_xSi_{20}$	0-10	SQ
$Pd_{80-x}Mn_xSi_{20}$	0-10	SQ
$Pd_{80-x}Fe_xSi_{20}$	0-10	SQ
$(Pd_{82}Si_{18})_{100-x}Fe_x$	10-90	SQ
$(Pd_{100-x}Fe_x)_{83,5}Si_{16,5}$	1-12	SQ
$Pd_{78}Si_{20}Fe_xCr_{2-x}$	0-2	SQ
$Pd_{80-x}Co_xSi_{20}$	0-10	P, SQ
$(Pd_{100-x}Co_x)_{83,5}Si_{16,5}$	0-18	SQ
$Pd_{80-x}Ni_xSi_{20}$	0-10	SQ
$(Pd_{100-x}Ni_x)_{83,5}Si_{16,5}$	0-60	SQ
$Pd_xCu_ySi_z$	x = 65-80 y = 3-19 z = 16-20,5	SQ

$(Pd_{82,4}Si_{17,6})_{100-x}Cu_x$	0-14	SQ
$(Pd_{100-x}Cu_x)_{83,5}Si_{16,5}$	0-26	SQ
$Pd_xAg_ySi_z$	x = 75-79 y = 4-8 z = 16-20	SQ
$(Pd_{100-x}Ag)_{83,5}Si_{16,5}$	0-18	SQ
$Pd_xAu_ySi_z$	x = 68-81 y = 4-12 z = 15-20	SQ
$(Pd_{100-x}Au_x)_{83,5}Si_{16,5}$	0-20 80-100	SQ
$(Pd_{100-x}Rh_x)_{83,5}Si_{16,5}$	0-6	SQ
$Pd_{84}Ge_xSi_y$	x = 2-7 y = 10-14 z = 77-79	SQ
$Pd_xAu_yAg_zSi_{16,5}$	y = 2-15 z = 3-4	SQ
$(Pd_{70}Mn_{30})_{100-x}P_x$	17-26	SQ
$Pd_{80-x}Fe_xP_{20}$	10-48	SQ
$(Pd_{100-x}Co_x)_{80}P_{20}$	15-60	SQ
$Pd_xNi_yP_z$	x = 0-82 y = 8-73 z = 10-23	SQ
$(Pd_{60-x}Pt_xNi_{40})_{75}P_{25}$	0-60	SQ
$Pt_{100-x}Ge_x$	17-30	SQ
$Pt_{100-x}Sb_x$	33-37	SQ
$Pt_{100-x}Si_z$	23,25,68	SQ
$(Pt_{100-x}Ni_x)_{80}P_{20}$	10-80	SQ
$(Pt_{100-x}Ni_x)_{75}P_{25}$	20-70	SQ
$(Pt_{70}Ni_{30-x}Cr_x)_{75}P_{25}$	1,5-6	SQ
$(Pt_{70}Ni_{30-x}V_x)_{75}P_{75}$	0-3	SQ
$Rh_{78}Si_{22}$	-	SQ
Rh-Ge	?	SQ
$Sn_{90}Cu_{10}$	-	SQ,EV

$Te_{100-x}Ga_x$	10-30	SQ
$Te_{100-x}Ge_x$	10-25	SQ
$Te_{100-x}In_x$	10-30	SQ
$Te_{70}Cu_{25}Au_5$	-	SQ
$Tl_{100-x}Te_x$	15-60	EV, SQ
$Tl_{100-x}Au_x$	25-60	SQ
$Zr_{72}Co_{28}$	-	SQ
$Zr_{100-x}Ni_x$	20-40	SQ
$Zr_{100-x}Cu_x$	40-75	SQ
$Zr_{100-x}Pd_x$	20-35	SQ
Y-Fe	?	SQ

EV = vapour deposition

ED = electrolyte deposition

CD = chemical deposition (electroless)

P = plasma-jet deposition

SQ = splat quenching

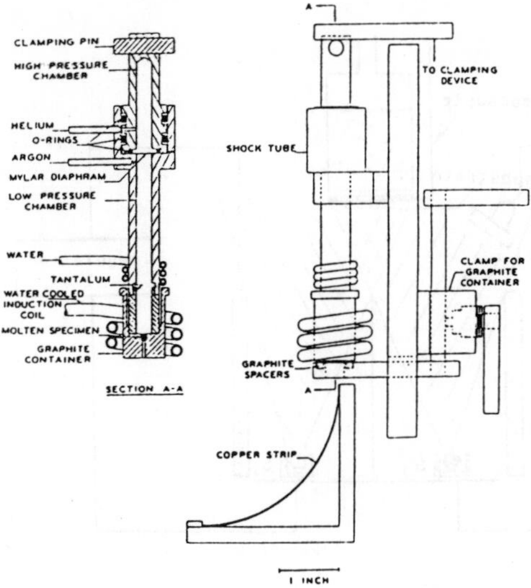

Fig. 1. Duwez apparatus for rapidly quenching of metals

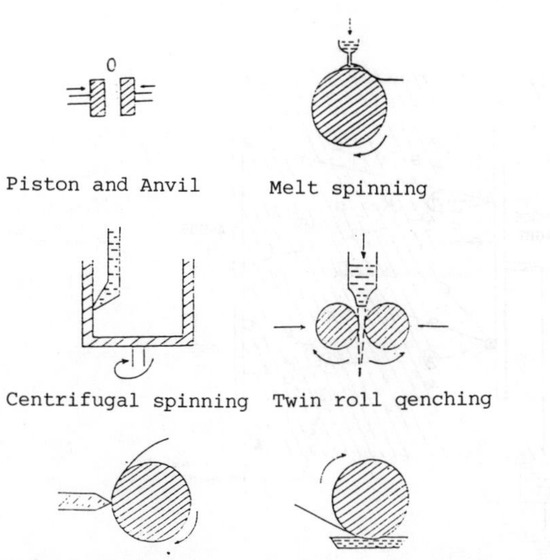

Fig. 2. Schematic representation of rapidly quenching processes

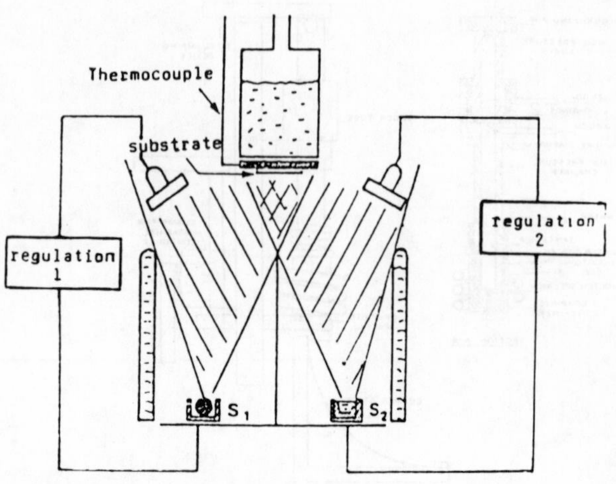

Fig. 3. Vacuum evaporation apparatus with two sources[14)]

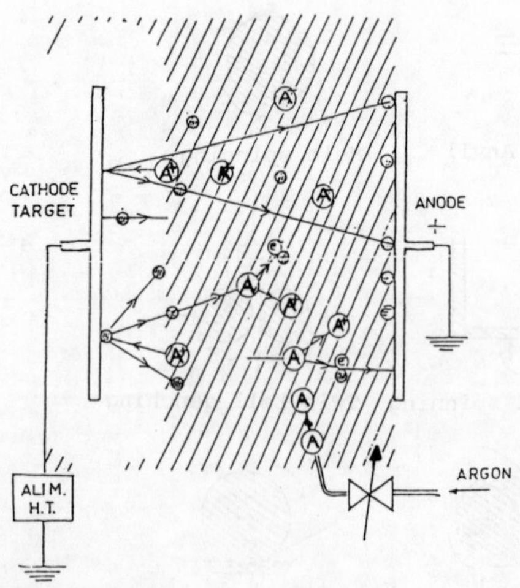

Fig. 4. Schematic representation of cathodic sputtering process[16)]

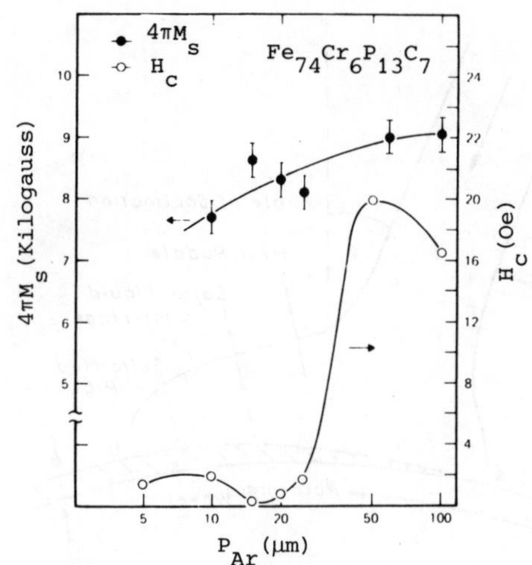

Fig. 5. Variation of the magnetization and of the coercitive force in function of the argon pressure for FeSiB sputtered samples[20)]

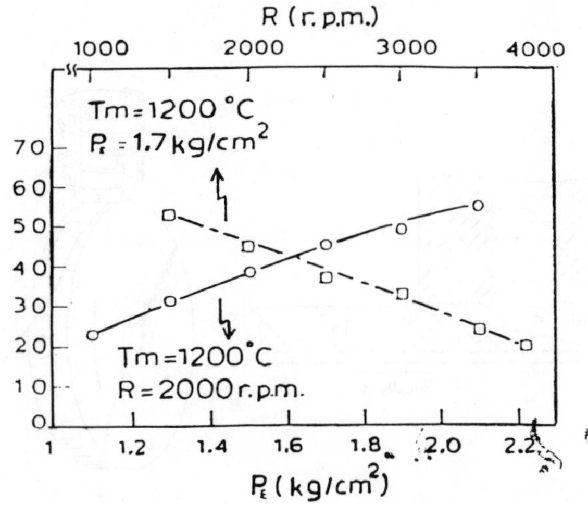

Fig. 6. Variation of the thickness of an amorphous ribbon with the ejection pressure and with angular speed of the wheel[30)]

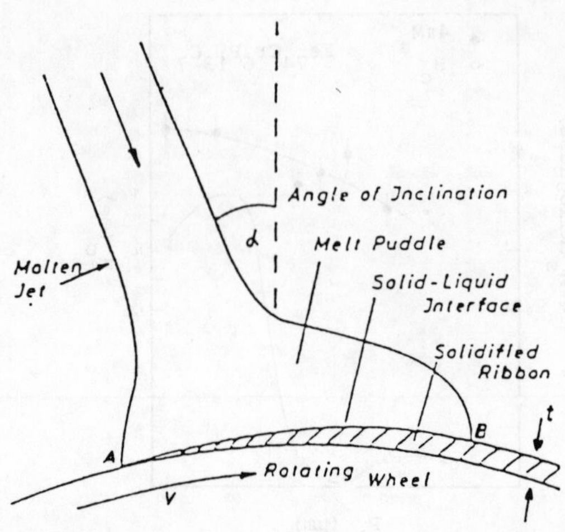

Fig. 7. Schematic representation of the formation of the ribbon in the melt spinning process[32)]

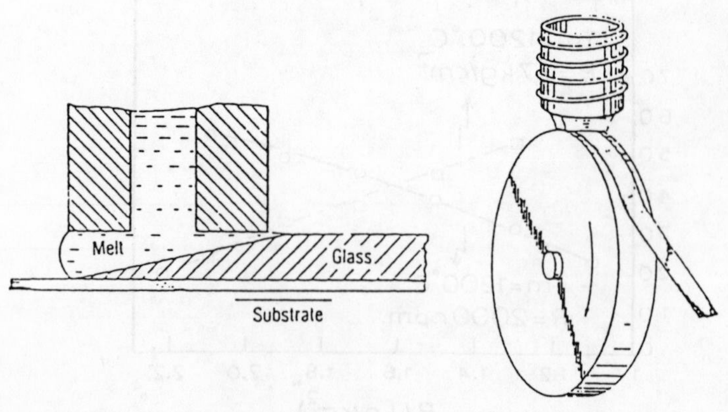

Fig. 8. Schematic view of the planar flow casting process[11)]

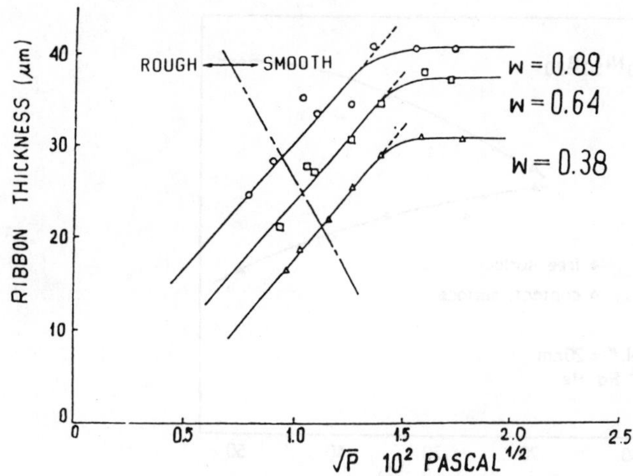

Fig. 9. Ribbon thickness variations as a function of the squere root of the ejection pressure with different slot breath w [40)]

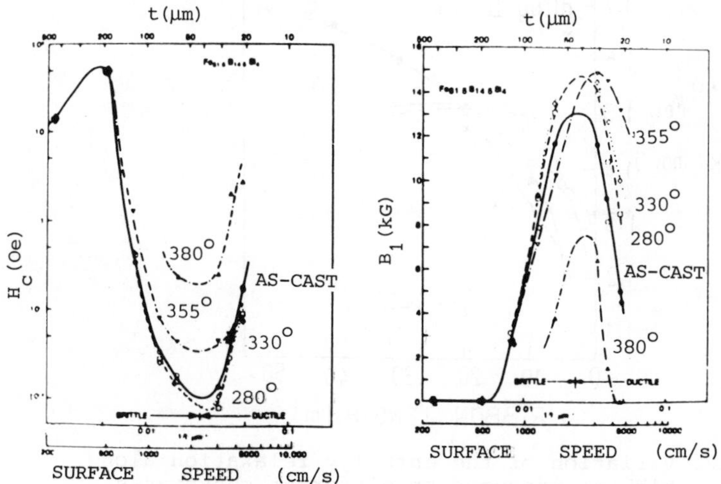

Fig. 10. The effect of surface speed of the substrate on the magnetic properties of amorphous ribbons[44)] ($Fe_{81,5}B_{14,5}Si_4$).

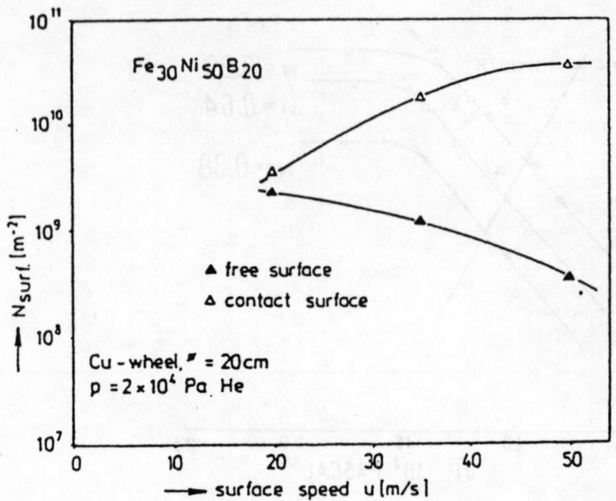

Fig. 11. Number of surface nucleation sites produced on the free side and on the contact side of the ribbons for different substrate velocities[50)]

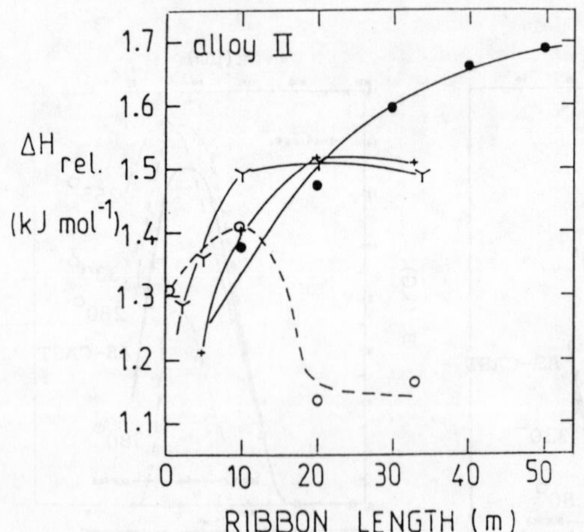

Fig. 12. Variation of the enthalpy relaxation along ribbons produced on different substrates with same sticking length[57)]
Wheel diameter 30 cm (●) Cu diameter 21 cm (+) Cu
(Y) CuBe
(o) soft iron

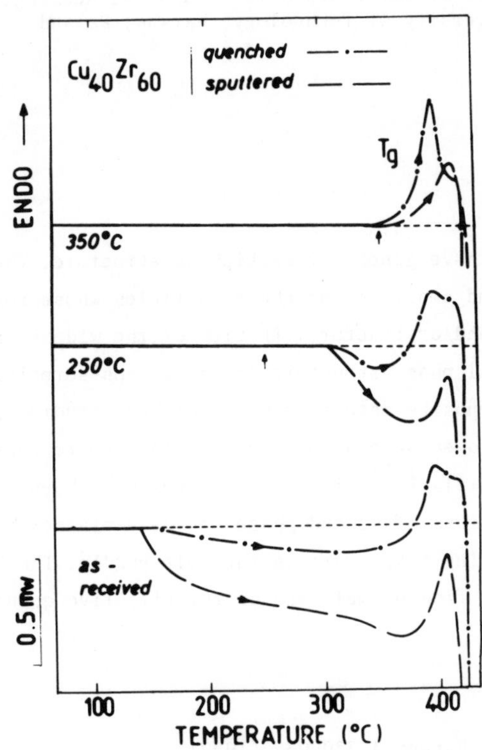

Fig. 13. Comparison of the DSC curves of the amorphous sputtered and liquid quenched $Cu_{40}Zr_{60}$ samples at different annealing temperatures [58]

RAPID SOLIDIFICATION OF COMPOSITE MATERIALS

Piotr G. Zielinski

Institute of Materials Science and Engineering,
University of Technology, Warsaw, Poland

Engineering materials have generally multiphase structure. The materials consist of a matrix and small second phase particles whose role is to control plastic flow and/or fracture. In view of the widespread technical application of dual phase structure it is perhaps surprising that rapidly solidified composite materials have not been studied in detail. Three different techniques have been developed to prepare rapidly solidified composite materials [1-3]. These techniques have been studied and the observed distributions of second phase can be explained by assuming that rotational stirring takes part in the melt puddle. The influence of second phase on the flow of melt and on the effective quenching rate will be discussed.

1) M.Narasimhan - U.S. Patent 4,330,037 (1982).

2) J.F. Williford, J.P. Pilger - U.S. Patent 3,776,297 (1973).

3) P.G. Zielinski, D.G. Ast - J.Mater.Sci.Lett. 2, 495 (1983).

PLASTIC DEFORMATION OF AMORPHOUS METALS

Piotr G. Zielinski

Institute of Materials Science and Engineering,
University of Technology, Warsaw,
Poland

The mechanical properties of amorphous metals are continuously investigated not only because these properties are important for potential application but also because the scientific explanations of these phenomena leave still open questions.

Plastic deformation of metallic glasses occurs by either of two mechanisms: a diffuse rearrangement of atoms at high temperatures or the nucleation and propagation of narrow shared regions. Three stages could be distinguished as the deformation proceeds at room temperature:

- a homogeneous stage, in which permanent plastic deformation occurs without the formation of shear bands,
- a second regime in which shear deformation proceeds by the formation and propagation of slip bands,
- a final stage in which deformation is concentrated in a few shear bands with crack formation along these bands.

The slip distribution along the shear bands can be analysed by modelling the band as a distributed pile-up of Volterra edge dislocations. The friction stress is about 0.8 of the critical shear stress in well developed bands. In composite materials with an amorphous matrix the shear bands cannot propagate past second phase particles. Both Young modulus and yield stress increase with the volume fraction of second phase particles. The increase in yield stress can be analysed with the one parameter work theory of Ashby. The value derived for the friction stress agree well with the value derived from analysis of bending in the plastic hinge regime and with the measurements of the slip distribution in shear bands. It is argued that the friction stress is closely linked to the proportional limit, i.e. the stress level at which local rearrangements first take place in a metallic glass. This conclusion is strongly supported by the observations of structural relaxation and embrittlement of Fe-Ni-B-Si metallic glasses. It was found that strain corresponding to the proportional limit increases after annealing in the temperature range in which partial annihilation of free volume occurs. The idea of small scale phase separation can be used for explanation of the embrittlement phenomena. If phase separation occurs the friction stress should change within the material affecting the propagation of the shear bands and leading to the catastrophic loss of ductility.

UNDERSTANDING THE STRUCTURE OF METALLIC GLASSES

P.H. Gaskell

University of Cambridge, Cavendish Laboratory
Madingley Road, Cambridge CB30HE, U.K.

1. INTRODUCTION

It is difficult to think of a more exciting time to be involved with research into the structure of amorphous solids. And perhaps the most interesting, most rapidly-developing field is that of amorphous metals. Two reasons for this statement can be given and both relate to the relatively short history of amorphous metals. Although it is over 30 years since the pioneering experiments of Werner Buckel on Sn-based amorphous thin films [1)] and the splat-quenching experiments of Pol Duwez and co-workers [2)], the expansion of interest in rapidly-quenched metallic alloys in the 1970's has had two important consequences.

a) We have learned more about the structure of certain metallic glasses over the short history of the subject than we have been able to deduce for most oxide glasses despite their longer history - dating back perhaps to Zachariasen's speculative comments in 1932 [3)] and Warren's subsequent elaboration by X-ray scattering techniques [4)]

b) Since the timescale for the development of knowledge of amorphous metals has been so compressed, it has been possible for individual researchers to watch the evolution of thought as it has happened - almost dynamically. This is in contrast to previous developments where ideas were propounded, elaborated and, sometimes, discarded by proselytising schools (often strongly nationalistic) over, perhaps, two decades. In the twelve years spanning my own interests in the structure of amorphous metals, it has been easy to see the progression of ideas from the minimalist notions associated with Dense-random-packing models of the mid 1970's through the structurally more complex ("Baroque", even) models involving some degree of local ordering, to the present tentative notions of a yet-to-be-established pattern of medium-range ordering.

As a final summary of the attraction of the subject in 1985, it is possible to argue that, despite all this effort, and notwithstanding the genuine advances which have taken place, the essential reason for glass formation still eludes us, as the ferment of current debate - involving substantial fundamental areas of disagreement between proponents of randomness, microcrystallinity, coherent domains, curved-space structures etc. - will testify.

In this paper I wish to present an attempt to record the evolution of a small part of the subject, glimpsed from one, arguably idiosyncratic, viewpoint, with no pretence that the conclusion will be final in any sense or that the picture is anything other than blurred. Much of the work derives from notions already published elsewhere: the inevitable result of a succession of fascinating conferences in a vintage year. The following might therefore be regarded less as an attempt at a critical essay but as a short story to be read on the train along with the crossword, the strip cartoons or the sports news, to be discarded and eventually (shortly) to be succeeded by more learned and lasting conclusions.

2. LOCAL STRUCTURE

For many amorphous solids the immediate environment of at least a subset of atoms has been defined and is known to be recognisably similar to coordination polyhedra in compositionally-equivalent crystals. This is certainly true for many covalent oxides and amorphous semiconductors - indeed it is difficult to think of one example where it has been proved that the local structure of a glass is random. For amorphous metals the situation is less clear: in certain cases, most notably the amorphous transition metal-metalloid glasses, a well-defined local coordination polyhedron appears to be almost beyond doubt. For many of the metal-metal glasses, although experimental evidence is not so extensive, even compositional ordering at the level of nearest-neighbours seems very doubtful. Largely because of the availability of good experimental data - but partly because the provisional conclusions are so challenging to our preconceptions of what glasses should be - the remainder of this article will focus on the transition metal-metalloid glasses and, as it transpires, the more structurally-ordered end of the spectrum.

2.1 Metalloid-Metalloid Avoidance

Maintaining the historical perspective: realisation of the inadequacy of the model of dense random packing of hard spheres (DRPHS) - which represented everyone's starting point in the early seventies - followed from the now classic work of Sadoc and Dixmier [5)] reported during the Second International Conference on Rapidly-Quenched Metals. Partial pair distribution function (PPDFs) derived for a-$Co_{81}P_{19}$, figure 1, clearly demonstrated the absence of P-P nearest neighbours at the distance anticipated from the atomic radius of the phosphorous atom. This lead to the concept of 'chemical short-range ordering' or 'metalloid avoidance' which became incorporated as a central tenet of DRP modelling of transition metal-metalloid alloys from then on, and has subsequently been shown to hold in all T-M glasses for which partials have been established and also in structurally similar alloys such as $Ni_{33}Y_{67}$ [6)]. Sadoc and Dixmier's results fitted in nicely with the concepts proposed earlier by Polk [7,8)] who suggested that Bernal's [9)] original DRPHS model for monoatomic liquid metals, subsequently extended by Finney [10)], might be modified for binary glasses by incorporating the smaller atom (metalloid in this case) within the larger interstices of the metallic DRP model. Metalloid-metalloid avoidance was thereby ensured with a neat synthesis of the ideas of random close-packing for the metal-metal interaction and some degree of selectivity for the metalloid environment. Polk was able to relate the fraction of large interstices to the composition (80:20) corresponding to easy glass formation in this system. Polk's model is, in my view, a powerfully attractive concept and, although not so much in evidence as in the past, may still be found to contain new essential grains of truth.

2.2. Metalloid Coordination Numbers

Sadoc and Dixmier's data indicated a coordination number of 8.9, suggestively close to the 9-fold coordination found in an overwhelming number of crystalline T-M alloys. Subsequently, partials were obtained for $Fe_{80}B_{20}$ [11], $Ni_{81}B_{19}$ [12] and $Ni_{80}P_{20}$ [13] by X-ray and neutron scattering and for $Pd_{78}Ge_{22}$ [14] and $Co_{80}P_{20}$ [15] by EXAFS techniques. In all cases, the immediate environment of the metalloid is found to be near the value of 9.0, table 1. (The four most accurate determinations provide a mean coordination number of 8.85 ± 0.09). The ratios of metalloid and metal atomic radii vary from 0.66 to 0.88 in these glasses and, if geometrical considerations and random packing alone were the determining factors, the coordination numbers should vary from about 7.5 to 9.0, figure 2.

2.3 The First Neighbour Bond Length Distributions

Accurate PPDF's allow the breadth of the M-M and T-M bond length distributions to be evaluated. These important parameters reflect the degree of distortion of the nearest-neighbour environment and give some clues about the arrangement of atoms in the coordination shell. Specifically, when experimental artefacts - broadening due to Fourier-transformation over truncated reciprocal space data and thermal vibrations - have been allowed for, the resulting distribution can be approximately characterized by the standard deviation, σ_s giving the static broadening due to deviations from a spherical first-neighbour shell. In all cases, σ_s for the metalloid-metal distance, M-T, is smaller than that for T-T, indicating a more ordered environment for the metalloid than for the transition metal. These results have, more recently, been confirmed measurement of the pressure-dependence of the hyperfine field gradient at the ^{57}Fe nucleus. The relative variation in the Fe-Fe distance is found to be about twice that for the B-Fe distance in a-Fe_3B [16] and is still greater in a-$(FeNi_3)_3B$ [17].

The M-T distribution also gives an indication of the "shape" of the nearest neighbour shell. An interesting feature of the M-T distributions for a-$Ni_{81}B_{19}$ [12], figure 3, and for $Pd_{84}Si_{16}$ [18] is the asymmetry: a small shoulder appears on the higher side. The same feature is seen in the crystalline T_3M structures (cementite lattice) and represents the distance from the central metalloid to three capping atoms. In contrast, those glasses and crystals which have larger metalloid atoms (Fe-B, Ni-P and Co-P), have a more regular distribution of M-T distances with little asymmetry, figure 4. In both cases the local structural unit in the crystal is a distorted capped trigonal prism - with distortions from the ideal structure increasing as the metalloid radius ratio increases and with a concomitant decrease in the difference between distances from metalloid to vertex and capping atoms. However, relative atom size is not the only determining factor - if so, the structures of Fe-B and Ni-B alloys would be almost identical as the radius of Fe and Ni atoms differ by only about 2%. The chemical nature of the elements apparently determines the different structures of the crystalline borides and, surprisingly, this seems to be reflected in the structure of the glasses and in their properties.

In the limit of small metalloid atoms and large metals the coordination shell of the metalloid will have minimal distortions and it then becomes possible to obtain fairly direct evidence of its shape. Alloys with a suitably small metalloid radius ratio are the Pd- [19] and Ru-Zr borides [20]. In both cases - especially the latter - it is possible to determine a second neighbour T-T distance corresponding to the diagonal of a square or rectangle of T atoms, figure 6, thus implying the presence of octahedral arrangements of the metal atoms, such as the square facets of the "crystalline" trigonal prism. The absence of such distances in amorphous monoatomic glasses and most binary alloys has long been considered powerful evidence for poly-tetrahedrality [21] as the essential ingredient of amorphous packing - a conclusion which almost certainly remains valid for the monoatomic alloys, but can now be seen to have an alternative explanation in binary alloys - namely distortion of a larger, more complex, local structural unit[22].

2.4 Symmetry of the Local Environment of the Metalloid

The symmetry of the local environment around a ^{11}B nucleus can be probed through the interactions between the nuclear quadrupole moment and the local electric field gradient. Pannisod and Bakonyi [23] have shown that the experimental spectra for a series of Ni-B alloys with B content in the range 18.5 - 40% is consistent with a model in which the environment of the B atom consists of a 9-atom tri-capped trigonal prism over the entire composition range. The Archimedean antiprism which might also represent a candidate polyhedron does not fit the data. These authors have examined the extent to which their data could be consistent with randomly packed spheres. Near the eutectic composition (18.5 at.%B) the environment of boron is well-defined - as revealed by the relative breadth of the electric field gradient distribution (σ/ν_Q) of about 0.15, which is lower than values predicted for the random arrangement (0.5). Values of σ/ν_Q increase as the boron concentration increases to about 30% when the random model cannot be excluded. However, the authors consider that trigonal prismatic symmetry is more appropriate throughout.

3. BEYOND THE LOCAL STRUCTURE

The results quoted above indicate that there are good reasons to consider the environment of the metalloid to be ordered - certainly a constant coordination number and relatively small fluctuations in the nearest neighbour M-T distances. The exact symmetry is open to a little more doubt but a capped trigonal prism appears to be the most appropriate coordination polyhedron figure 5.

The notion that the local environment in the glass is similar to that of the corresponding crystalline phase is not new. This is the common supposition for most covalent glasses, and has been suggested as a guide for amorphous metals by many workers - for example, Maitrepierre [24] and Polk [7,8]. The idea gained added impetus from the Mössbauer work of the Hungarian school who coined the term 'quasicrystalline' model to express the similarity of the assumed local structure to that of the crystal [25].

Around the same time, I described the properties of an essentially speculative model for a-T-M alloys [26-28] in which the immediate environment of the metalloid is defined as a 6-atom trigonal prism, but, in addition, specific rules are built in which describe the connectivity of one unit to the next (thereby fixing the position of a seventh 'capping' atom). Specifically, a series of trigonal prisms were connected according to the edge-sharing arrangement observed in the cementite (Fe_3C) structure, but with a random choice of the shared edge, figure 5, constrained only by the need to suppress void formation and thus maximize the model density. Radial distribution functions and structure factors computed for this type od model were in reasonable agreement with the experimental data available at that time, figure 7. Subsequently, experimental evidence has tended to support the central tenet of this model - a well-defined local structure unit in the form of a distorted trigonal prism - as described in section 2.

However, subsequent experimental evidence has cast doubt on two of the other tenets of this model: a) the notion that trigonal prisms are packed randomly and b) the rules for connecting trigonal prisms.

Partial distribution functions for $Ni_{81}B_{19}$, figure 3, [13], show that B-B correlations are still recognisable beyond 1.2 nm. If the boron lies in the centre of a trigonal prism, then this implies that the trigonal prisms are packed so that they are correlated to third or fourth neighbours - which is certainly not the case in the randomly packed trigonal prism model. Secondly, the interconnection between prisms apparently needs to be modified to produce a split first neighbour B-B peak. One possibility is that sheets of prisms connected as in cementite, are separated by an additional plane of close-packed metal atoms, in the process converting the composition from T_3M, corresponding to cementite (Fe_3C), to T_4M - approximately the composition of the glass [29].

In addition, an explanation for the existence of trigonal prisms is not straight-forward. As mentioned earlier, the model represented a speculative proposal - an attempt to carry the analogy of well-defined local structure from oxides, chalcogenides, and amorphous semiconductors into the field of amorphous metals. For covalent structures there is not the same problem. Well-defined trigonal and tetrahedral units are the hallmark of strong directional bonding from the central atom. In the case of Ni-B alloys it is difficult to see how boron atoms, say, with no d-electrons, can have bonds to nine first neighbours which are anything other than strong and non-directional. There is therefore no good reason for imagining that the T_9M coordination polyhedron has the character of a 'molecule' - to be clipped together to form a network. This view is reinforced by simple calculations [30)] which show that the trigonal prism has, if anything, rather less stability than another candidate polyhedron - the bi-capped (10-atom) Archimedean antiprism.

An alternative view is that the trigonal prism is formed as an essential part of an extensive medium-range structure [28)] in the same way as tetrahedra and octahedra form the primitive units of close-packed structures even though the octahedron does not necessarily represent a minimum energy for the corresponding isolated clusters. According to this view, the facts presented in section 2 do not merely constitute evidence in favour of a particular local coordination polyhedron but support the existence of certain medium-range structures which are built by packed arrangements of such units. In other words, the ordered local structure may be a consequence of an ordered medium-range structure.

What the preferred medium-range structure might be is, in my view, one of the most important outstanding questions. Several lines of approach may be considered: we could divide the structures which might form by efficient packing of trigonal prisms into three sub-sets.

a) Structures similar to crystalline solids with periodicity, but only over domains of appropriate size (1-2 nm), preferably with well-defined boundaries to other domains. (An alternative microcrystallite structure - with ill-defined interfaces is unlikely to survive even very rapid quenching from the melt without some evidence of grain growth triggered by energetic interface states).

b) Structures based on aperiodic packing or 'non-crystallographic' packing. Such structures - for example, polytetrahedral arrangements - have a long history as candidates for liquids and glasses. They may represent stable arrangements of small numbers of atoms but ultimately become relatively unstable due to their inability to pack 3-D Euclidean space thus leading to defects or strain.

c) Random arrangements which may be preferred over alternatives by virtue of the large number of combinations of the constituent polyhedra which can lead to efficient, space-filling packing and could therefore be considered to be favoured entropically.

At the moment, models have been devised and evaluated for the first and third category. Specifically, Dubois, Le Caer and Gaskell [29] have considered a model based on chemical twinning which extends the packing principles observed in crystals to the amorphous state. The essential notion is that a glass consists of domains within which atomic positions are correlated in a manner similar to compositionally equivalent crystals. Specifically, crystal structures corresponding to T-M alloys can be constructed by admitting twinning planes separating close-packed planes of metal atoms, and if such twinning planes are oriented along {112} of the h.c.p. structure, sheets of trigonal prisms are found which provide suitable cavities for metalloid atoms. By varying the spacing of such 'chemical' twinning planes, the composition of the alloy may also be varied and for the 80:20 (T_4M) composition, a 'chemical' twinning plane must be inserted every three {112} planes.

The model for an amorphous phase presupposes domains of about 1.0 - 1.5 nm within which all twinning planes are parallel. In neighbouring domains, the same principles hold except that the orientations of the twinning planes are different. They are not randomly orientated, however - a further condition imposed in this model is that atoms lying in the boundary regions between domains shall be subject to the constraints imposed by the chemical twinning of each of the domains meeting at that interface. Such atoms are, therefore, highly constrained and only a limited range of mutual orientations of neighbouring domains allows a sufficiently low energy boundary.

In this regard the model differs essentially from microcrystallite models. The results are encouraging. For example, figure 8 shows the computed partial pair distribution functions for a-$Ni_{81}B_{19}$ together with experimental data. A respectable fit is seen for distances up to about 0.6 nm.

In addition, Beyer and Hoheisel [31)] have used a molecular dynamics simulation technique to build an (ostensibly) random model for Ni-B and other metallic glasses. This model and that of Dubois et al fit the experimental data for amorphous nickel-boron about equally well (and equally badly at high r-values).

To date, although models have been proposed for T-M glasses which are based on aperiodic packing of trigonal prisms [32)] none has been evaluated in detail.

4. FINALE

Since the intention, in writing this article, has been to present a quasi-historical view of one aspect of this subject, from the early 1970's to the present, it would clearly be inappropriate to write a "conclusion". The sequel must follow - and indeed is already being planned. However, it is not inappropriate to remark that the final punctuation mark to the subject at this time - a query (?) - also applies outside the narrow boundaries of transition metal-metalloid glasses. Local structural order has been recognised in other unlikely glasses - for example, the environment of Na and Ca in silicate glasses appears to be relatively well-defined [33,34)] - a fact which might not have been suspected from the Zachariasen model for oxide glasses. In such cases where the coordination number is high and where local coordination polyhedra are extensively coupled to their neighbours by shared atoms, for instance, then local and medium-range structures are almost inextricably woven. In these cases, where the local structure is found by experiment to be well-defined, then a well-defined medium-range structure must be suspected. Certainly, the choice of medium-range structures is by no means completely free but must, to a greater or lesser extent, be constrained by the fact that the local structure is well-defined. The importance of such constraints remains to be established but it might not be unreasonable to suppose that in some amorphous solids - especially those near the edge of a glass-forming region, the net effect is that quite complicated medium-range structures must be generated before local coordination requirements can be adequately satisfied.

Acknowledgement

Generous support from Pilkington Brothers PLC. is gratefully acknowledged.

5. REFERENCES

1) Buckel, W., N. Physik, 138, 136 (1954).
2) Klement, W., Willens, R.H. and Duwez, P. Nature (London) 187, 869 (1960).
3) Zachariasen, W.H., J.Am Chem.Soc., 54, 3841 (1932).
4) Warren, B.E., Krutter, H. and Morningstar, O., J.Am.Cer.Soc. 21, 49 (1938).
5) Sadoc, J.F. and Diximier,J., Mater.Sci.Eng. 23, 187 (1976).
6) Maret, M., Chieux, P., Hicter, P., Atzmon, M. and Johnson, W.L. in 'Rapidly Quenched Metals' (eds. Steeb, S. and Warlimont, H. North-Holland, Amsterdam). Vol. 1, 521 (1985).
7) Polk, D.E., Scripta Met., 4, 117 (1970).
8) Polk, D.E., Acta Metall., 20, 485 (1972).
9) Bernal, J.D., Proc.Roy.Soc. A280, 299 (1964).
10) Finney, J.L., Proc.Roy.Soc. A319, 479 (1970).
11) Nold. E., Lamparter, P., Olbrich, H., Rainer-Harbach, A. and Steeb, S., Z. Naturforsch, 33a, 327 (1978).
12) Lamparter, P., Sperl, W., Steeb, S. and Bletry, J., Z. Naturforsch, 37a, 1223 (1982).
13) Lamparter, P. and Steeb, S., in 'Rapidly Quenched Metals' (eds. Steeb, S. and Warlimont, H., North-Holland, Amsterdam), Vol. 1,

459 (1985).

14) Hayes, T.M., Allen, J.W., Tauc, J., Giessen, B.C. and Hauser, J.J., Phys.Rev.Lett. 40, 1282 (1978).

15) Flank, A.M., Lagarde, P., Raoux, D., Rivory, J. and Sadoc, A., in 'Rapidly Quenched Metals IV (eds. Masumoto, T. and Suzuki, K.: Jap. Inst.Metals, Sendai), 393 (1982).

16) Abd-Elmeguid, M.M., Micklitz, H. and Vincze, I., Phys.Rev. B25, 1, (1982).

17) Eickelmann, H., Abd-Elmeguid, M.M., Micklitz, H. and Brand, R.A., Phys.Rev., B29, 2443 (1984).

18. Sadoc, J.F. and Dixmier, J., in 'The Structure of Non-Crystalline Materials' (ed. Gaskell, P.H. : Taylor and Francis, London), p.85, (1977).

19) Cocco, G., Enzo, S., Sampoli, M. and Schiffini, L., J.Non-Cryst. Solids, 61 and 62, 557 (1984).

20) Mehra, M., Williams, A. and Johnson, W.L., Phys.Rev. B28, 624 (1983).

21) Cargill G.S. in Solid State Physics (eds. Ehrenreich, H., Seitz, F. and Turnbull, D. : Academic Press, New York) 227, (1975).

22. Gaskell, P.H., in 'Rapidly Quenched Metals IV' (eds. Masumoto, T. and Suzuki, K.: Jap.Ins.Metals, Sendai) Vol. 1, 247 (1982).

23. Pannisod, P., Bakonyi, I. and Hasegawa, R., Phys.Rev. B28, 2374 (1983).

24. Maitrepierre, P.L., J.App.Phys. 44, 1189 (1973).

25. Kemeny, T., Vincze, I., Fogarassy, B. and Arajs, S. in 'Rapidly Quenched Metals III' (ed. Cantor, B. : Metals Soc., London), 291 (1978)

26. Gaskell, P.H., ibid. p. 27 (1978).

27. Gaskell, P.H., Nature 276, 484 (1978).

28. Gaskell, P.H., J.Non-Cryst.Solids 32, 207 (1979).

29. Dubois, J.M., Gaskell, P.H. and Le Caer, G., Proc.Roy.Soc. A402, 323 (1985).

30) Gaskell, P.H., in 'Rapidly Quenched Metals', (eds. Steeb, S. and Warlimont, H. : North-Holland, Amsterdam) Vol. 1, 413 (1985).

31) Fukunaga, T. and Suzuki, K., Sci.Rep.Res.Inst.Tohoku Univ. A29, 153 (1981).

Table 1

Experimental mean r_1, standard deviation, $\sigma_s(r_1)$ and coordination number, N, for nearest-neighbour M-T and T-T distribution in several amorphous alloys. The standard deviation, σ_s, represents an approximation to a (static) disorder-induced broadening of the first neighbour distribution since termination broadening has been subtracted. Thermal broadening is still included, however.

Alloy	Atom pair	r_1/nm.	$\sigma_s(r_1)$/nm	N	Ref
$Co_{81}P_{19}$	P-Co	0.232	0.011	8.9 ± 0.5	5
	Co-Co	0.254	0.016	10.0 ± 0.4	
$Fe_{80}B_{20}$	B-Fe	0.214	0.010	8.6	11
	Fe-Fe	0.257	0.017	12.4	
$Ni_{81}B_{21}$	B-Ni	0.211	0.014	8.9	12
	Ni-Ni	0.252	0.015	10.5	
$Ni_{80}P_{20}$	P-Ni	0.228	0.012	9.3	13
	Ni-Ni	0.256	0.017	9.4	
$Pd_{80}Ge_{20}$	Ge-Pd	0.249	0.01	8.6 ± 0.5	14
$Co_{80}P_{20}$	P-Co	0.22	≃0.01	9	15

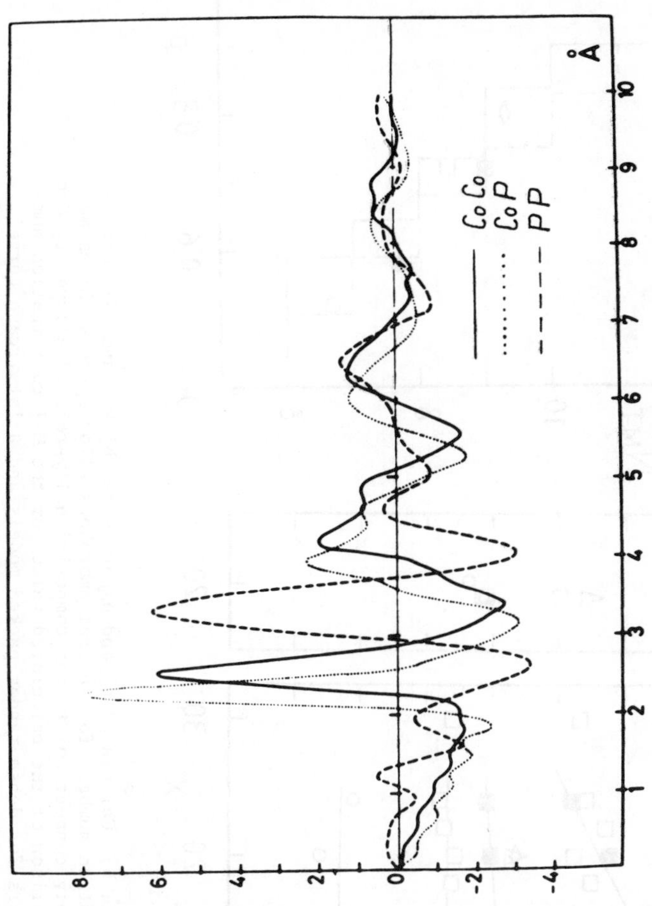

Fig.1. Partial pair distribution functions for a-$Co_{81}P_{19}$ deduced from X-ray and polarised neutron scattering measurements by Sadoc and Dixmier[5]. This work showed clearly, for perhaps the first time, the absence of next-neighbour phosphorous atoms which would be expected near 0.2 nm. The small peak observed near this distance is an artefact of Fourier transformation from reciprocal space data.

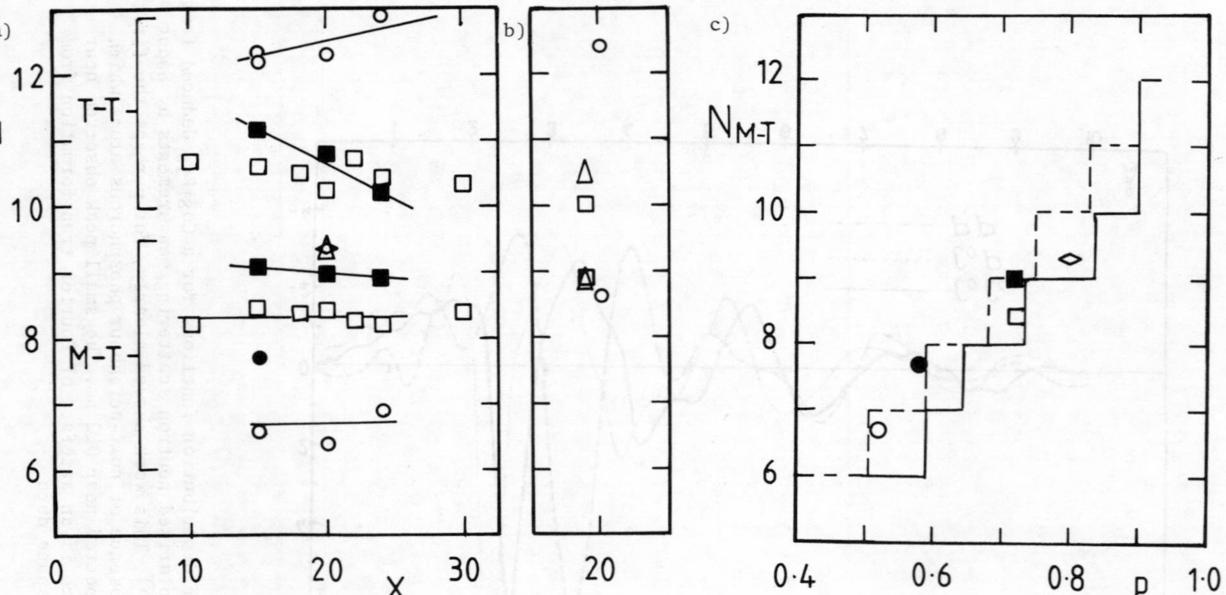

Fig.2 a), b). Calculated a) and experimental b) M-T and T-T coordination numbers for several amorphous alloy systems. Lines merely connect points for compositionally-related alloys. c) Comparison of the calculated values of the M-T coordination numbers for dense random packed models and a 'staircase curve' giving the number of spheres surrounding a central sphere of radius ratio p. For details see Gaskell 6).

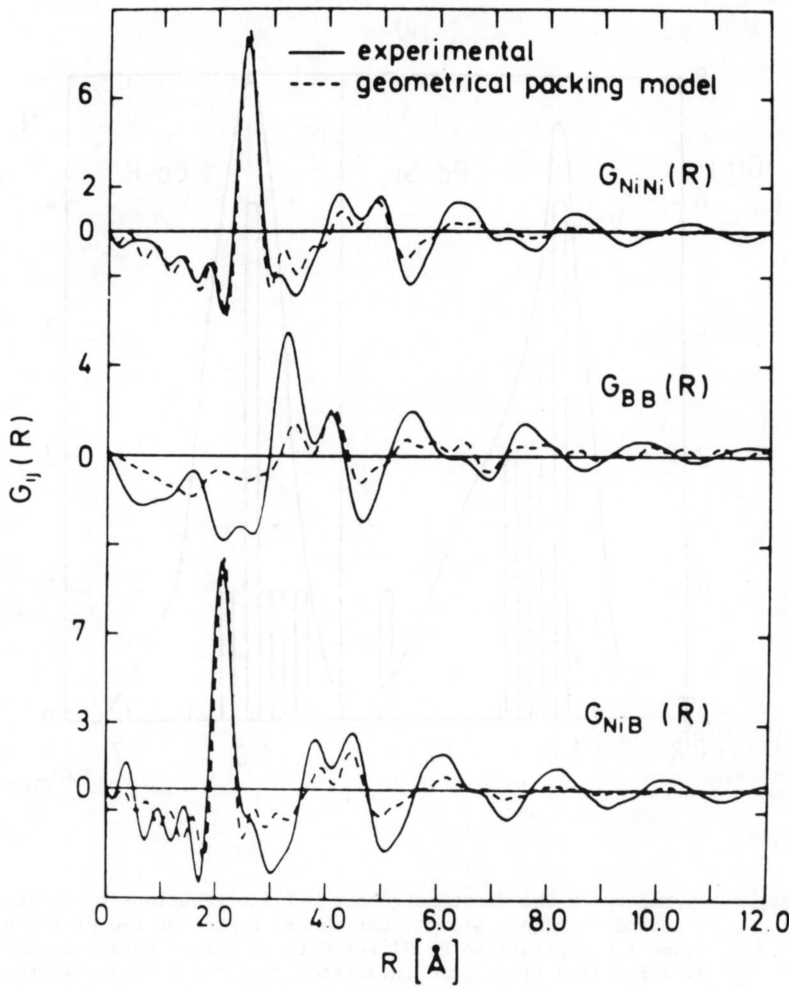

Fig.3. Partial pair distribution functions for a-$Ni_{81}B_{19}$ obtained by neutron-scattering measurements from specimens containing different isotopes of Ni [12]).

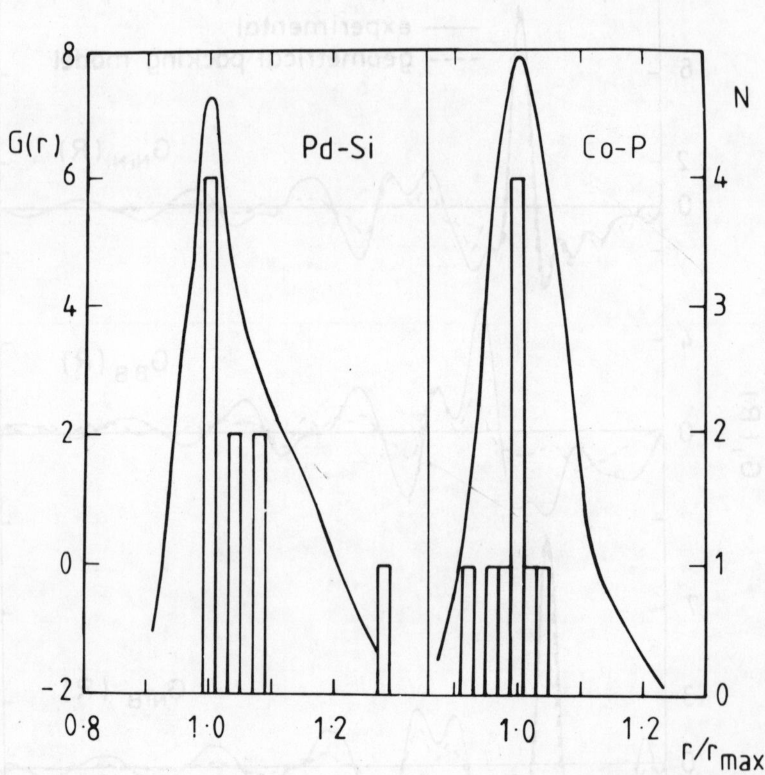

Fig.4. Asymmetry in the first-neighbour M-T distributions in crystalline and amorphous alloys. Bar charts represent the distributions for crystalline Pd_3Si (Fe_3C-type) (right hand ordinate) as a function of reduced inter-atomic distance r/r_{max} where r_{max} corresponds to the peak of the distribution. Curves give the corresponding distribution functions for Si-Pd and P-Co (left hand ordinate) in a-$Pd_{84}Si_{16}$ and $Co_{81}P_{19}$. Note the asymmetry of the distributions for crystalline and amorphous Pd-Si alloys, and similar asymmetry in figure 3. The distributions for a-$Co_{81}P_{19}$ and the Fe_3P lattice are more symmetrical.

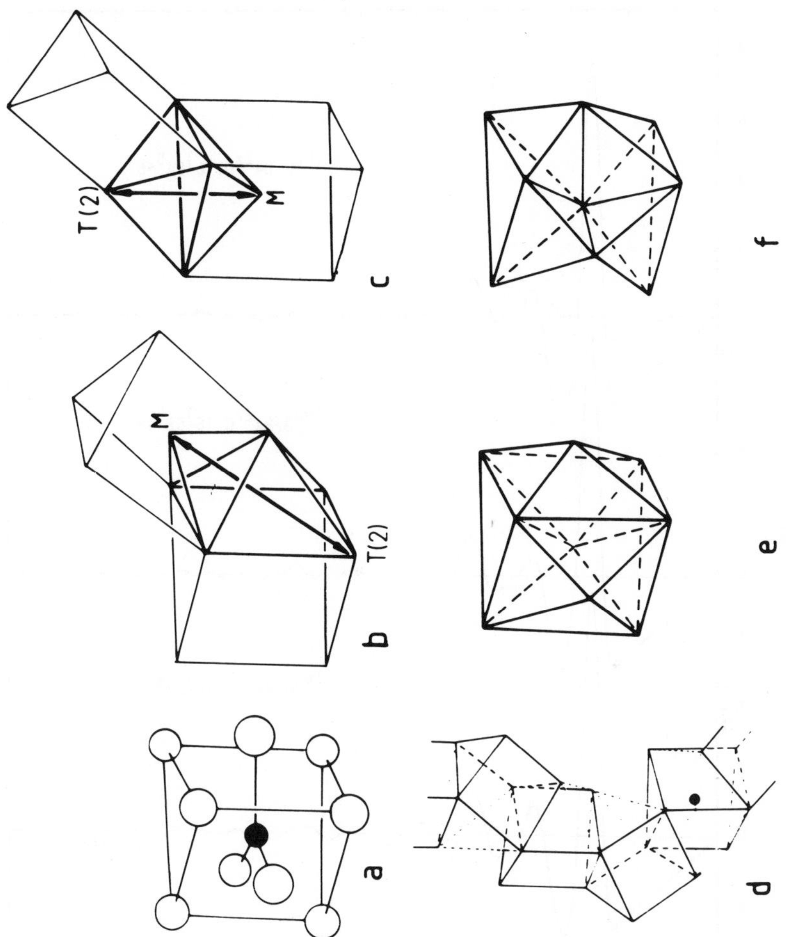

Fig.5. a) Trigonal prism with capping atoms. b), c). Two prisms with a common edge linked as in the Fe_3C lattice and Fe_3P respectively. d) Disordered arrangement of prisms produced by random edge sharing. e), f). Distorted prisms appropriate for alloys with small and large radius ratios respectively and their disection into tetrahedra or octahedra as appropriate.

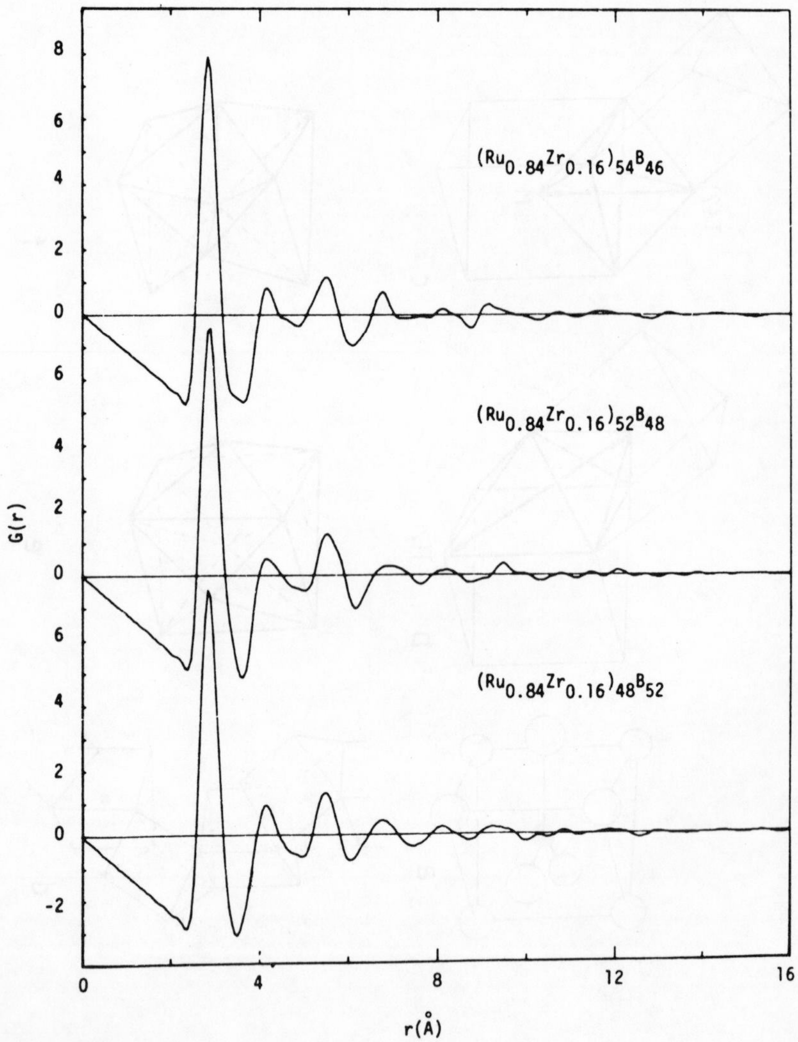

Fig.6. Reduced radial distribution functions, G(r), for three Ru-Zr-B alloys [21]. Note the prominent second peak near 0.4 nm. corresponding to distances of 1.46 times the first-neighbour T-T peak - representing (probably) the diagonals of octahedra.

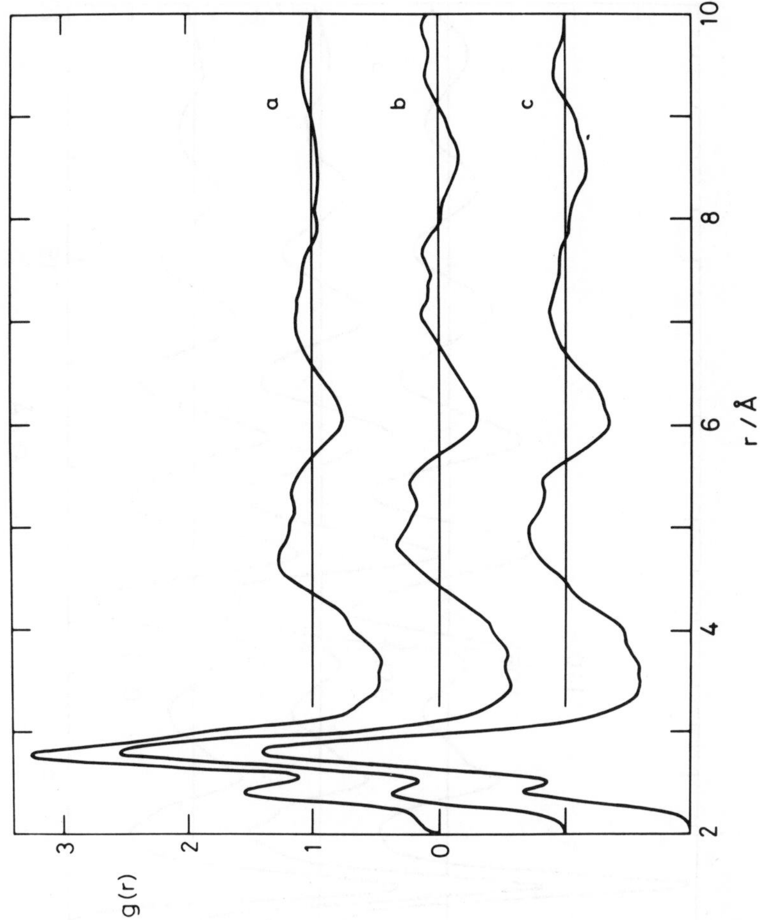

Fig.7a) Pair correlation functions, $g(r) = \rho(r)\rho_0$ where $\rho(r)$ is the atom density, and ρ_0 the averaged atomic density for a-$Pd_{80}Si_{20}$ [31] - curve a. Computed values of g(r) for two models based on random packing of trigonal prisms - curves b and c. For details see Gaskell [26-28].

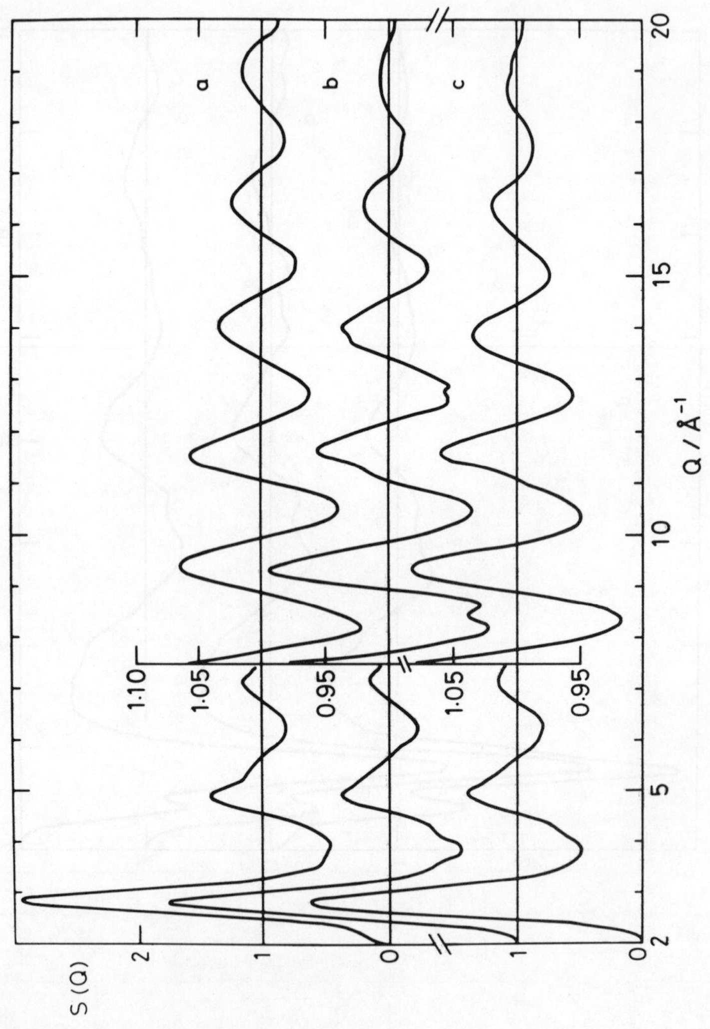

Fig.7b) Corresponding data for the structure factor, S(Q) of $Pd_{80}Si_{20}$.

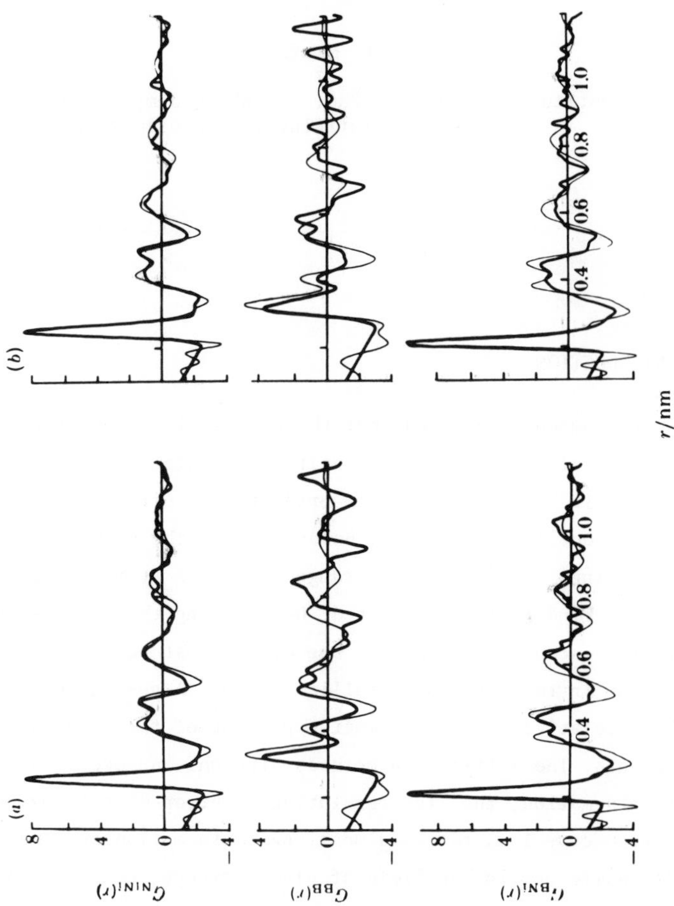

Fig.8. Partial pair distribution functions for two models of a-Ni_4B compared with experimental data of Lamparter et al [12]. The models have a domain size of 1.2 µm. For details see Dubois et al [29]. Note the improved fit compared with the random atomic packing model shown in Figure 3. Full line corresponds to calculated data.

COMPUTER MODELLING OF THE STRUCTURE OF METALLIC GLASSES

P. Mrafko

Institute of Physics EPRC, Slovak Academy of Sciences,
842 28 Bratislava, Czechoslovakia

1. INTRODUCTION

It is commonly accepted that there is no long-range order in the structure of glass. Knowing the positions of atoms in one region of an amorphous solid is not helpful in predicting their positions in a distant region. At short range, however, over the distance of a few atomic spacings and over medium range up to ten atomic spacings an order of some kind is to be expected. Determining the best way to describe this microscopic structure is one of the most active areas of investigation in the study of metallic glasses. A general problem concerns the relatively low information content of the usual structural measurements. The diffraction methods give only an averaged one-dimensional information, the high resolution electron microscopy and ion-field microscopy lead to two-dimensional information and the local methods, quite new in the field of glass science, namely, EXAFS, NMR, Mössbauer and Raman spectroscopy provide ways of probing an environment of atoms in the structure and the symmetry of atomic sites only.

A three-dimensional information, i.e. a full representation of disordered structure in real-space, can be obtained only by means of modelling. Therefore, it is important to complement experimental structure information by developing model structures which are built by hand or more comfortably by computer. The structure results from the competition between local interactions, which tend to arrange atoms in specific relation with neighbouring atoms, and topological and geometrical restrictions imposed by the space filling requirements. The ultimate goal is, of course, to find direct relation between atomic structure and various physical properties.

To start with the construction of a model we need firstly to strip the interatomic potential function down to those aspects which we think are structure-determining and secondly to devise a method of building the array of a large number of simplified atoms which is not constrained by conditions of periodicity in three-dimensional space.

2. INTERATOMIC POTENTIAL FUNCTIONS

2.1 Hard Potential

Obviously, no real atom can be represented as a hard sphere. Nonetheless, dense random packing of hard spheres are the most highly developed models for these materials. It is not only because of their simplicity. There are strong geometrical reasons, geometrical factors governing solid state structure [1)]. It is the tendency of "good" space-filling (space principle), the tendency to form arrangements of high symmetry, not only crystallographic (symmetry principle), and the tendency of atoms to form connections of high dimensions (connection principle). Of course, there are counteracting factors such as the formation of special bonds or the effect of temperature. Moreover, the hard sphere structures are extremly well-defined, the packing algorithms are strictly deterministic and the models are reproducible and easily analysed.

2.2 Soft Potentials

Single-component amorphous structures depend upon the potential and this dependence is probably even stronger for the binary metallic alloys. In metals we cannot ignore many-body effects from the delocalized electrons and hence density-dependent potential must be used. Moreover, interatomic potentials were found to vary strongly with composition due to charge transfer [2]. Generally the interatomic potentials are unknown. Intuitively we suppose that a "good" potential will lead naturally to a good representation of the structure. The situation is not so simple since we have to find an adequate method of soft sphere packing as well.

3. METHODS OF MODELS BUILDING

Historical development leads to three conceptual models of structure of amorphous metals. These are microcrystalline, random packing and the so-called designed or stereochemically-defined models. The proper representation of the structure of amorphous state will be probably created by some mish-mash of these concepts.

3.1 Microcrystalline Models

The structure factors for many amorphous metals and alloys show some similarities with data for corresponding crystalline alloys. The idea is to represent the structure of disordered phase as an assembly of microcrystals. Disorder results from small sizes of the well-defined crystals, their random orientation, grain boundaries and defects within microcrystals. We can use the Debye formulat to calculate structure factors, however, it is important to include the interference between microcrystallites as well [3]. Generally this kind of models was not succesfull, the main reason being probably high energy associated with interfaces between crystallites.

3.2 Random Packing Models

3.2.1 Static methods

A. Hard sphere packing

i) Sequential method. This method consists of sequential or serial deposition of atoms onto a seed. We start with a seed usually in the form of a tetrahedra. There are vaious possibilities for choosing the site for a new atom on the surface. In the case of binary alloy the control of alloy composition is also not trivial. One cannot simply decide which species to add by use of a random number generator because each addition adjusts the composition of the model only localy. The composition has to be recomputed for each trial site [4]. There are problems with the homogeneity and isotropy of the model created as well.

ii) Non-sequential methods. Initial coordinates of atoms can be chosen completely at random and the atom radius increases from zero until overlaps occur when either the overlapping atom is removed or each atom is displaced to remove the overlap.
"Gas compression" algorithm - N points are generated at random and we suppose them to be centres of spheres. Overlapping spheres are moved apart along their line of centres until they just touch, ignoring any new overlaps created. When all overlaps have been removed, the sphere radius is increased and the process repeated. Clusters created are isotropic and homogeneous [5].

B. Soft spheres packing

We use some starting set of points as centres of atoms (randomly created or from hard sphere packing) and we try to find non-crystalline structure with minimum energy. There is a possibility for a dependence of the structure on starting configuration. This dependence will increase for multicomponent system. In addition to the packing constrains that dominate the single component structure we must consider the "dispersion" of the two components in the alloy.

We are looking for the steepest-gradient path on the energy-hypersurface in the multidimensional configuration space. An alternative is the steepest-descent algorithm, the displacement of individual atoms by amounts proportional to the resultant force and in the direction of the resultant force. We have no guarantee that the minimum energy structure obtained by such iterative processes represents global minimum. This can be turned to advantage as the order of initial structure is in some way preserved so we do not loose at least some degree of control over the character of the model. A better reproducibility of minimalization is obtained by "annealing" by means of mechanical cycle, e.g. a compression strain is applied step-by-step up to, let's say, 10%, then at each step the dimension of the cluster is decreased by 1% and the energy is minimized with respect to the atomic coordinates. Then the strain is gradually decreased using the same procedure. Finally, full minimization is made with respect to coordinates and density variation.

3.2.2 Dynamic methods. In the molecular dynamics (MD) method the atoms are each assigned a velocity at random, ensuring that the average kinetic energy corresponds to the required temperature [6]. The forces acting on each atom are calculated and the atoms are allowed to move according to Newton's laws. Temporal development follows (and the required temperature adjusted by changing velocities of atoms) until the assembly is equilibrated; the structural and dynamic properties can be extracted.

The Monte Carlo (MC) method works rather differently but produces sample configurations which are in the limit consistent with MD. Starting with the configuration of the required density, a randomly chosen atom is moved in a random direction by a random amount and the potential energy change is calculated. If it is negative, the motion of the atom is accepted and the process repeated; otherwise, the motion is accepted with a Boltzmann weighting [7].

We can follow the evolution of the system in time, we can calculate some time-averaged properties or we can produce snapshot of atomic positions. Long relaxation times of non-crystalline or glassy materials, compared with the elementary MD time step, presently limit full simulation of quenching rates to much higher than those achieved in laborarory. MC suffers from similar problems in that the temperature must be reduced (or pressure increased) unrealistically rapidly. It might be argued that for much slower cooling rates the atoms might be able to explore phase space more thoroughly before falling into to final position. MD and MC methods produce a configurationally-arrested liquid which will necessarily ignore those rearangements that will occur on the time-scale of the laboratory quench.

The combination of static and dynamic methods looks very promising in the sense that after a change of temperature a few steps of MD are carried out followed by energy minimization combined with mechanical annealing. Finally, reaching required equilibrium state, full minimization is made with respect to coordinates and density variations.

3.3 Designed Structures

The main problem in modelling of binary alloys is the feeding-in of local constrains and local chemical order (if we now what it is) at each stage during construction of the structure model.

It is found that some of the many groupings of atoms possess advantageous energetic or space-filling properties. It may be reasonable to assume that such groups will predominate in a given structure. It is thus possible to create essentially random-packing model with a high probability of non-crystallographic groups of atoms, e.g. starting from a random arrangements of atoms we can relax step by step a small compact group of atoms (e.g. 13) taken randomly. The procedure converges and the final structure will contain much more (distorted) units (e.g. ikosahedra) then the fully relaxed one [8)].

Another useful working rule may be that the nearest-neighbour coordination of certain atoms in glass is almost identical with that observed in the corresponding crystals. The idea is firstly to build-in the desired chemical ordering (perhaps to overemphasize it) and then to allow the relaxation process to bring in the dense packing constrains using the model potential function. This function is designed to favour the original localy-ordered structure. Gaskell [9)] first explored the possibilities of the method for metal-metalloid systems where trigonal prismatic coordination seems to be probable. Model was built initially in the laboratory (later it was transfered to the computer) by joining trigonal prismatic units in a special way. The model was then carefully relaxed with some constrains, e.g. in one type of model the second neighbour metal atom is treated differently to try to emphasize the trigonal prismatic ordering in the relaxed structure. Thus a structure is introduced in which the order extends to the nearest-neighbour shell, and perhaps beyond it. The experimental observations suggest correlated rather than random (at least in some alloys) arrangements of local structural units. The nature of the medium-range structure, that is, the mode of packing of local structural units is more difficult to forecast.

The model based on chemical twinning [10] can be considered as some extension of this basic idea of trigonal prismatic coordination in the direction of correlated arrangement of local structural units. The model is based on the concept that domains of positionally-correlated atoms (basically arrays of trigonal prisms) exist on a scale 1-2 nm. Within each domain the positional order is governed by rules appropiate to a single type of structural operation - simple chemical twinning in this case. Positions of atoms in the interface between domains are not chosen arbitrarily but similar constraints are used as within the domains in order to reduce stress and strain so that this starting structure has little in common with microcrystalline models. Steepest descent algorithm is applied to reduce total energy.

4. INFLUENCE OF BOUNDARY CONDITIONS

Every model must be capable of extension indefinitely in three dimensions without any significant change in the structure, i.e. the model boundary must be consistent with the constraints that would be imposed by the presence of similar arrangement of atoms at that boundary.

In the case of finite clusters, some thousand atoms are typically used. In the cluster of this size more than 30% of the atoms are located at the surface. Their near-neighbour statistic and the local number density are quite different from those of the bulk and the unbalancing of the interatomic forces at the surface will introduce considerable distorsion in the cluster. Periodic boundary conditions are used; they probably do not affect pair correlation function, but no complex microstructural analysis has been performed. Thus it is possible that the periodic replication of the cluster introduces some elements of lattice symmetry into the results.

5. ANALYSIS OF MODELS

In the case of crystalline solids we have no problems in specifying structures in detail; however, the lack of adequate characterisation of non-crystalline systems makes difficult the comparisons among amorphous structures. We need much more effective ways of describing structures which will focus on the significant microstructural differences.

Given a set of atomic positions, it is then possible to calculate several microscopic properties such as packing fraction or density, number of near-neighbours, structure factor and pair correlation function (PCF). One of the basic problems in the theory of amorphous solids is the description of their structure beyond the level of the exclusive use of PCFs. Although the information contained in a set of PCFs is sufficient to express many bulk properties such as the equation of state and even to describe chemical short-range order, it is clear that it is inadequately sensitive to the topological short-range order and is insuficient to describe properties which depend critically one the local functions of structural parameters.

Voronoi polyhedra are used for analysis of the environment of atoms, i.e. the local structure found around individual atoms. A Voronoi polyhedron is a polyhedron formed around a centre of atom by planes drawn to bisect perpendicularly vectors drawn between all pairs of centres of atoms in the model. Each polyhedron contains all the points closer to the central atom than to any other. For binary packing the radical plane construction is preferred [11)], the dividing plane between different atoms is drawn at distances proportional to the atomic radii. The volume of Voronoi polyhedron corresponds to the volume of an atomic cell and its distribution is a natural measure of the structural randomness. The microstructure of the model may be represented by statistic of this polyhedra.

By defining near-neighbour distance (e.g. as the first minimum of PCF) it is possible to decompose the structure into tetrahedra, octahedra and various types of deltahedra (as icosahedra) with atoms at vertices and to look for a particular arrangement of these [12)].

Cavities inside these bodies as well as their spatial distribution can be calculated.

The local environment of atoms can be characterized by local levels of stress and strain or by local fluctuation of density. In addition the local atomic symmetry coeficients are defined which indicate the deviation from spherical symmetry around the atom [13].

6. CONCLUSION

Knowledge of microstructure is a basic assumption for building up a general theory of glassy state and the modelling is one of the most promising ways to gain this knowledge. So far the experiments and structural modelling do not uniquely define a real amorphous structure; however, the proposed models have been successfully used and some of them are actually consistent with the experiments for "reasonable" values of their parameters. Data from these models will certainly play an important role in the development of adequate theories.

REFERENCES

1) Laves, F., Internmetallic Compounds ed. J.H. Westbrook, Chapt.8, John Willey Comp. 129, 1967.

2) Gurman, S.J., Proc. V Int. Conf. on Rapidly Quench Metals , eds. S.Steeb and H. Warlimont (Elsevier Sci. Publ. B.V., 557 1985).

3) Hamada, T., Fujita, F.; Jap. Journ. Appl. Phys. 21, 981 (1982).

4) Boudreaux, D.S., Gregor, J.M., J. Appl. Phys. 48, 5057 (1977).

5) Mrafko, P., J. de Phys., Colloque C8, 41, 322 (1980).

6) Raman, A., Mandel, M.J., McTague, J.P., J. Chem. Phys. 64, 1564 (1976).

7) Krajci, M., Mrafko, P., J. Phys. F: Met. Phys. 14, 1325 (1984).

8) Mrafko, P., Proc. 8 Konf. Czech. Physicist, Bratislava, 1-200, 1985.

9) Gaskell, P.H., Proc. III Int. Conf. on Rapidly Quench Metals (ed. B. Cantor) Metals Soc., London, Vol. 2, 277 (1979).

lo) Dubois, J.M., Gaskell, P.H., Le Caer, G., submitted to Phil. Mag.

11) Gellatly, B.J., Finney, J., J. Non-Cryst. Sol. 50, 313 (1982).

12) Lancon, F., Billard, L., Chamberod, A., J. Phys. F: Met. Phys. 14, 579 (1984).

13) Egami, T., Maeda, K., Vitek, V., Phil. Mag. A41, 883 (1980).

ATOMIC TRANSPORT IN AMORPHOUS METALS

A.L. Greer

University of Cambridge
Department of Metallurgy and Materials Science
Pembroke Street, Cambridge, CB2 3QZ, U.K.

1. INTRODUCTION

On deliberate annealing, or under service conditions, amorphous metals may undergo crystallisation, structural relaxation or phase separation. Since atomic rearrangement is involved in each of these processes, direct studies of atomic transport (i.e. diffusion and viscous flow) may prove useful in increasing our understanding both of the stability of amorphous alloys and of heat treatments to develop optimal properties. We will consider mainly diffusion, and in order to be relevant to structural changes, we will be concerned only with the diffusion of components which might be themselves constituents of amorphous alloys. Specifically excluded is a treatment of hydrogen diffusion, which has been much studied in amorphous alloys and displays many interesting phenomena [1-3].

2. MEASUREMENT OF ATOMIC DIFFUSIVITY

Measurement of atomic diffusion in amorphous alloys is difficult because of their lack of stability; to avoid crystallisation, the measurement must be made at a low temperature, certainly not exceeding the glass transition temperature T_g, and diffusivities are consequently low, typically 10^{-17} to 10^{-24} m^2s^{-1}. Such diffusivities can be measured by sensitive profiling techniques. The diffusing component is added to the specimen surface by coating [4] or by ion-implantation [5] and after diffusion anneals the composition profiles are measured by sputter-sectioning combined with analysis by Auger electron spectroscopy [6], secondary ion mass spectrometry [4], or radiotracer [7]. To avoid the destruction of the sample necessary in sputter sectioning, the analysis may be performed by Rutherford backscattering [5] or by using a nuclear reaction analysis [8]. These profiling techniques have a depth resolution of about 3 nm and a minimum measurable diffusivity of about 10^{-24} m^2s^{-1}. This sensitivity is sufficient to avoid crystallisation, but not structural relaxation, during diffusion anneals. If it is assumed that the length of diffusional jumps is approximately the atomic diameter, then in a typical anneal necessary to give a measurable composition profile, an atom makes at least 200 jumps [9]. Thus the glass is expected to become fully relaxed during the diffusion anneal, and though there is some disagreement [5], it indeed appears that prior relaxation has little effect on diffusivities measured by profiling methods. Also, it should be noted that samples produced by melt-spinning may be substantially relaxed because of the relatively slow cooling after the ribbon leaves the wheel [10].

Methods other than profiling techniques have also been used to determine the low diffusivities in amorphous alloys. Köster et al.[11] have measured the diffusion-controlled growth rate of crystals in a variety of transition metal-metalloid glasses. The method is sensitive and accurate, but the diffusion coefficient can be identified only if the rate-controlling process is known. It seems, however, that metalloid diffusion is rate-controlling; certainly the data on an Fe-Ni-B glass are in excellent agreement (fig.1, [12]) with diffusion measurements from profiling [13] and from boriding and deboriding experiments [14]. This chemical method is limited in application (tough perhaps adaptable to other elements), but very sensitive.

3. GENERAL RESULTS

Diffusivities have been measured in several amorphous alloys, and a review of the results has been given by Cantor and Cahn [15].

Here only the major, general, features will be noted.

Within the relatively narrow temperature range in which it is possible to make measurements, and within experimental error, all the diffusivities in amorphous alloys show an Arrhenius temperature dependence. The activation energies vary from 145 to 350 $kJmol^{-1}$. The frequency factors have values between 10^{-12} and 10^{12} m^2s^{-1}; such a wide range of values is not physically reasonable, and it is probable that no significance can be attached to values obtained by fitting the behaviour in a narrow temperature range. Figure 2 shows diffusivities of a variety of solutes in amorphous systems. It seems clear that the coefficients at a given temperature decrease in the order: interstitial diffusion in crystals, grain boundary diffusion, diffusion in amorphous alloys, substitutional diffusion in crystals. The very clear distinction between interstitial and substitutional transport in crystals, with rates differing by 6 to 15 orders of magnitude, does not exist for amorphous alloys.

In comparing behaviour of different amorphous alloys it is usual to normalise the temperature scale using the glass transition temperature. When this is done, fig.3 shows that smaller atoms do diffuse somewhat faster. This ordering of the data suggests that T_g is a suitable normalising parameter. Cantor and Cahn [15)] reported that normalisation of the diffusivities by dividing by $a^2\nu$ (a ... atomic diameter, ν ... is the atomic vibration frequency = kT_g/h) is not significant as $a^2\nu$ varies only in the range 3×10^6 to 5×10^6 m^2s^{-1} for all the amorphous alloys studied. Again on a normalised temperature scale, a given atom, for example gold as pointed out by Cantor [16)], diffuses at similar rates in a wide variety of metal-metalloid glasses and at similar rates in a variety of metal-metal glasses, with the coefficients in the latter case being about 2 orders of magnitude greater.

The effect of amorphicity itself on diffusion coefficients could be assessed only by comparing crystalline and amorphous systems of the same composition. Unfortunately, diffusion coefficients in crystalline systems suitable for this comparison have not been measured. One reported value, for the diffusion of carbon in Fe_3C, is very low and does suggest [9)] that in crystalline metal-metalloid systems of glass-forming composition the bulk diffusivities of even small atoms may be at least as low as in the amorphous alloys, and not comparable to interstitial diffusivities in the crystalline metals.

4. MECHANISMS OF ATOMIC TRANSPORT

In attempting to understand the mechanisms of atomic transport in amorphous alloys it is useful to compare atomic diffusion and viscous flow, and to consider structural relaxation effects.

Viscous flow can be monitored very accurately in creep experiments. Taub and Spaepen [17] found that, due to structural relaxation, the viscosity of $Pd_{82}Si_{18}$ glass rose linearly with isothermal annealing time, and it was possible for the viscosity to reach at least 10^5 times its original value. This marked dependence of the viscosity on the degree of relaxation renders difficult the determination of an activation energy for viscous flow, but Taub and Spaepen showed that in well-annealed samples the fractional change in viscosity during a determination was negligible. By performing successive anneals at different temperatures and then verifying that the viscosity at the original temperature had not changed significantly, it was possible to make a comparison of the amorphous alloy at one degree of relaxation and hence determine the <u>isoconfigurational</u> activation energy; it was found to be 192 kJ mol^{-1}. That it is possible to measure viscous flow without the structure relaxing perceptibly, need not be surprising when it is realised that in a sensitive creep experiment only 1 atom in 10^4 need jump in a viscosity determination.

Structural relaxation effects on the rates of atomic diffusion can be detected reliably only by using techniques more sensitive than those described in Section 2. One possibility is the use of artificially layered amorphous alloys. These consist of alternating thin (typically 1-2 nm) layers of two compositions prepared, for example, by sputtering [18]. On annealing, the layers interdiffuse and the Bragg satellite peak due to the composition modulation decays. By monitoring the decay, interdiffusivities $\tilde{D}$ as low as 10^{-27} m^2s^{-1} can be measured. Using this technique on a multilayer with compositions $Fe_{85}B_{15}$ and

$Pd_{85}Si_{15}$, Greer et al.[19] monitored $\tilde{D}$ during an anneal and found that structural relaxation did have an effect, causing $\tilde{D}$ to decrease by at least 2 orders of magnitude. This type of experiment is still less sensitive than a creep test, but only 0.2 jumps per atom are required for a determination of $\tilde{D}$. It was found that $1/\tilde{D}$ rose linearly with isothermal annealing time, and that the isoconfigurational activation energy for diffusion (determined in the same way as that for viscosity) was 195 kJ mol^{-1}. It seems that in the examples cited there is a close link between atomic diffusion and viscous flow: $1/\tilde{D}$ is proportional to the viscosity η, and the isoconfigurational activation energies are the same.

Such a close relationship is to be expected on the basis of the free volume theory for transport in liquids. In this theory, atomic rearrangement of the shearing type shown in fig.4 can occur only at a site where there is sufficient local free volume (which in a glass may be considered a defect in the ideal amorphous structure). Such local shearing leads to both diffusion and viscous flow. The rates of the two processes should be proportional, and for liquid metals are found[20] to be related by the Stokes-Einstein equation:

$$\eta D = \frac{kT}{6\pi r} , \qquad (1)$$

where k is Boltzmann's constant, T the absolute temperature and r a characteristic distance, given approximately by the ionic radius. For the amorphous alloys cited above it is possible to compare the linear rates of increase of $1/\tilde{D}$ and η, and thereby to avoid the difficulty of comparing different samples with the same degree of relaxation. It is found[9] that the measured diffusivity is at least 150 times the value that would be predicted from the viscosity using eq. (1). Therefore it seems that there may be diffusive jumps which do not contribute to viscous flow. Crystalline metals offer a good example of this: in Nabarro-Herring creep, vacancy flow contributes to viscous flow only at sites where the vacancies can be created or destroyed. The comparison of diffusive and viscous flow rates has been for amorphous alloys of the metal-metalloid type. In these alloys the local atomic configurations appear to be well defined[21], and it may be possible to have a defect in the local order, analogous to the vacancy in a crystal, which is relatively stable and primarily responsible for diffusive transport.

Tsao and Spaepen [22] performed viscosity measurements on $Pd_{77.5}Cu_6Si_{16.5}$, an amorphous alloy remarkably resistant to crystallisation. They found that eventually the linear rise in viscosity slows down as equilibrium is approached. If viscous flow is assumed to occur at particular defects, then the population of defects, n, decreases with bimolecular kinetics:

$$\dot{n} \propto -(n-n_e)^2 , \tag{2}$$

where $\dot{n} \equiv dn/dt$ and n_e is the equilibrium population of defects at the annealing temperature. By first annealing for a long time at a low temperature, it was found that an anneal at a higher temperature could be accompanied by a viscosity <u>decrease</u>, as n increased towards the new higher value of n_e. Within the experimental error, the kinetics of the viscosity decrease could be bimolecular or unimolecular [22]. It is interesting to note that in conventional silicate glasses the kinetics of viscosity increase are bimolecular and of viscosity decrease are unimolecular. These two phenomena arise from the joining of dangling bonds and bond breaking respectively. That the behaviour of amorphous alloys is similar is perhaps evidence for a well-defined chemical ordering similar to that in silicates.

If diffusion occurs predominantly at faults in the amorphous structure this would explain simply the lack of a strong distinction between interstitial and substitutional diffusion. The importance of the chemical bonding particularly is emphasised by the low diffusivity found for carbon in crystalline Fe_3C (Section 3).

5. DIFFUSION BARRIERS

It is appropriate to point out that the special properties of amorphous alloys - lack of fast grain boundary and interstitial diffusion - may be of some technological importance. Refractory amorphous alloys have been tested as diffusion barriers under the metal contacts on semiconductor devices [23]. As barriers the alloys are excellent until the onset of crystallisation, and the resistance to crystallisation appears to be adequate. The electrical characteristics in Schottky diodes and in ohmic contacts are good. One problem is that low diffusivity in the amorphous layer does not prevent diffusion of the components of the layer into the overlying metal or the underlying semiconductor, and examples of both have been found.

6. SOLID STATE AMORPHISATION

Diffusion studies in amorphous alloys have recently acquired greater importance because of the demonstration that amorphous alloys can be formed by solid state interdiffusion of crystalline elemental layers. The principle was first demonstrated by Schwarz and Johnson [24], who showed that alternating thin (10 to 60 nm) layers of gold and lanthanum would interdiffuse at 50 to 80°C to form an amorphous alloy. The free energy of the amorphous alloy, though somewhat higher than would be achieved by the formation of crystalline intermetallic compounds, is much lower than that of the layered mixture of elements. The nucleation of the crystalline compounds is kinetically suppressed at the low annealing temperature. The thin layers of two metals can be produced not only by thin film deposition, but also by the progressive cold deformation of an initial macroscopic composite, and in this way it is possible to make "bulk" amorphous samples. For example, Schultz [25] has used wire drawing of a Ni-Zr composite to produce 0.8 mm diameter wires which after annealing were fully amorphous. Atzmon et al. [26] have reported the formation of 200 μm thick amorphous foils of Ni-Er and Cu-Er by cold rolling and annealing.

Originally it was suggested [24] that the amorphisation reaction was due to anomalously fast diffusion of one element in the other (in this case of gold in lanthanum). However, for the Zr-Co system Schröder et al. [27] have shown, by transmission electron microscopy of sections cut across the layers, that the amorphous phase grows as planar layers between the elemental layers. This finding, which is likely to be generally true, shows that diffusive transport in the amorphous phase is essential for the reaction and almost certainly rate-controlling. Schröder et al. found that Kirkendall-type voids were formed because the cobalt was diffusing faster than the zirconium in the amorphous layers. This asymmetry of diffusion behaviour has been found in other amorphising systems, including Ni-Zr, where elegant marker experiments [28] have shown that the nickel diffuses at least ten times faster than the zirconium. Spaepen [29] has suggested that this asymmetry of diffusion behaviour is essential for amorphisation: the fast transport of one species enables homogenisation, while the slow transport of the other prevents nucleation of the stable crystalline intermetallics. The asymmetry of behaviour in the amorphous phase may be related to anomalous fast diffusion of one element in the other in the crystalline state. It was pointed out above that there is no evidence for a distinction between substitutional and interstitial diffusion in amorphous alloys. Nonetheless, diffusion coefficients for different species in a given amorphous alloy can very by about two orders of magnitude, and this is sufficient to accommodate the asymmetry found in the amorphisation experiments.

7. AREAS FOR FURTHER WORK

For diffusion in amorphous alloys it is possible to identify some areas in which further investigation would be of particular interest.

1. Work is needed on the composition-dependence of diffusivities in amorphous alloys. To do this work, experiments will have to be devised in which the total composition range is small. This range is certainly not small in present profiling or multilayer experiments.

2. The effect of structural variations, arising from different preparation methods and heat treatments, should be more thoroughly tested. A direct comparison of diffusion rates in melt-spun and sputter-deposited materials would be particularly useful.

3. There is much more scope for measurements of self-diffusion using isotopes.

4. Measurements over wider temperature ranges are needed to provide a better test of Arrhenius behaviour.

5. It would be worthwhile to relate direct diffusion measurements (e.g. by profiling) to the kinetics of solid state amorphisation, in order to identify any special features of the processes in that reaction.

ACKNOWLEDGEMENTS

The author's research in this area is supported by the Science and Engineering Research Council (U.K.), the Office of Naval Research (U.S.A.), the International Business Machines Corporation (U.S.A.), and Plessey Research, Caswell, Ltd. (U.K.).

REFERENCES

1) Berry, B.S. and Pritchet, W.C., Phys. Rev. B24, 2299 (1981).
2) Bowman, R.C. and Maeland, A.J., Phys. Rev. B24, 2328 (1981).
3) Kirchheim, R., Sommer, F. and Schluckebier, G., Acta Metall. 30, 1059 (1982).
4) Cahn, R.W., Evetts, J.E., Patterson, J., Somekh, R.E. and Kenway-Jackson, C., J. Mater. Sci. 15, 702 (1980).
5) Chen, H.S., Kimerling, L.C., Poate, J.M. and Brown, W.L., Appl. Phys. Lett. 32, 461 (1978).
6) Luborsky, F.E. and Bacon, F. in Proc. 4th Int. Conf. on Rapidly Quenched Metals, Sendai, 1981 (eds. T. Masumoto and K. Suzuki) p. 561. Japan Institute of Metals, Sendai (1982).
7) Gupta, D., Tu, K.N. and Asai, K.W., Phys. Rev. Lett. 35, 796 (1975).
8) Birac, C. and Lesueur, L., Phys. Status Solidi A36, 247 (1976).
9) Greer, A.L., J. Non-Cryst. Solids 61 62, 737 (1984).
10) Akhtar, D. and Misra, R.D.K., Scripta Metall. 20, 627 (1986).
11) Köster, U., Herold, U., Hillenbrand, H.-G. and Denis, J., J. Mater. Sci. 15, 2125 (1980).
12) Köster, U., private communication.
13) Kijek, M., Ahmadzadeh, M., Cantor, B. and Cahn, R.W., Scripta Metall. 14, 1337)1980).
14) Brodowsky, H. and Sagunski, H., Z. Phys. Chem. Neue Folge 139, S.149-152 (1984).

15) Cantor, B. and Cahn, R.W., in Amorphous Metallic Alloys (ed. F.E. Luborsky) ch.25. Butterworths, London (1983).

16) Cantor, B., in Proc. 5th Int. Conf. on Rapidly Quenched Metals, Würzburg, 1984 (eds. S. Steeb and H. Warlimont) p.595 North Holland, Amsterdam (1985).

17) Taub, A.I. and Spaepen, F., Acta Metall. 28, 1781 (1980).

18) Rosenblum, M.P., Spaepen, F. and Turnbull, D., Appl. Phys. Lett. 37, 184 (1980).

19) Greer, A.L., Lin, C.J. and Spaepen, F., in Proc. 4th Int. Conf. on Rapidly Quenched Metals, Sendai, 1981 (eds. T. Masumoto and K. Suzuki) p. 561. Japan Institute of Metals, Sendai (1982).

20) Nachtrieb, N.H., in Liquid Metals and Solidification, p.49. American Society for Metals, Cleveland, OH (1958).

21) Gaskell, P.H., in Glass - Current Issues (eds. A.F. Wright and J. Dupuy) p.54. Martinus Nijhoff, Dordrecht (1985).

22) Tsao, S.S. and Spaepen, F., Acta Metall. 33, 881 (1985).

23) Nicolet, M.-A., Suni, I. and Finetti, M., Solid State Tech. 26, 129 (1983).

24) Schwarz, R.B. and Johnson, W.L., Phys. Rev. Lett. 51, 415 (1983).

25) Schultz, L., in Proc. 5th Int. Conf. on Rapidly Quenched Metals, Würzburg, 1984 (eds. S. Steeb and H. Warlimont), p. 1585. North Holland, Amsterdam (1985).

26) Atzmon, M., Unruh, K.M. and Johnson, W.L., J. Appl. Phys. 58, 3865 (1985).

27) Schröder, H., Samwer, K. and Köster, U., Phys. Rev. Lett. 54, 197 (1985).

28) Cheng, Y.-T., Johnson, W.L. and Nicolet, M.-A., Appl. Phys. Lett. 47, 800 (1985).

29) Spaepen, F., in Proc. Int. Cryogenic Materials Conf., Cambridge, Mass., 1985 (eds. T. Orlando and S. Foner). Plenum, New York, in press.

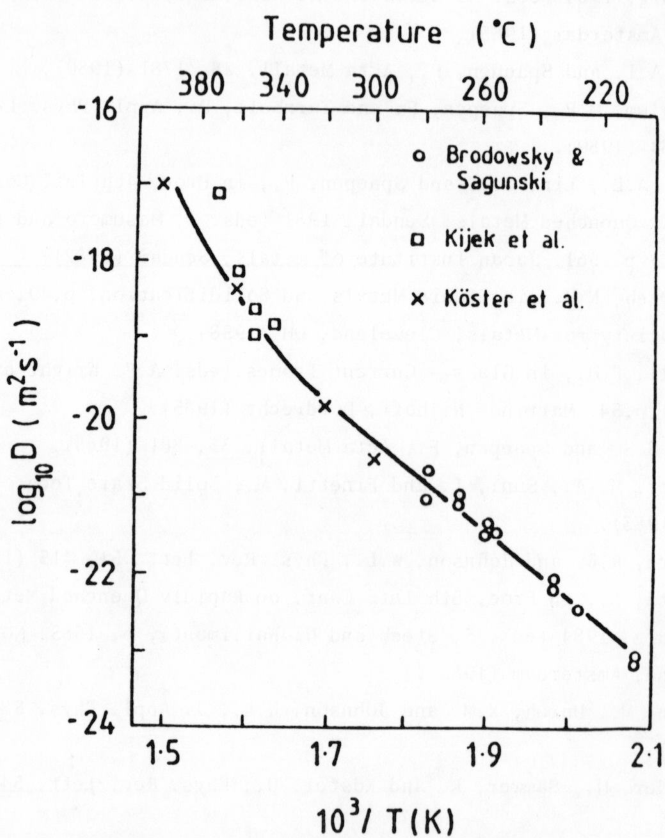

Fig.1. The diffusivity of boron in Fe-Ni-B amorphous alloys as measured by boriding and deboriding (Brodowsky and Sagunski [14]), profiling by ion-milling and SIMS (Kijek et al. [13]), and primary crystallisation measurements (Köster et al. [11]).

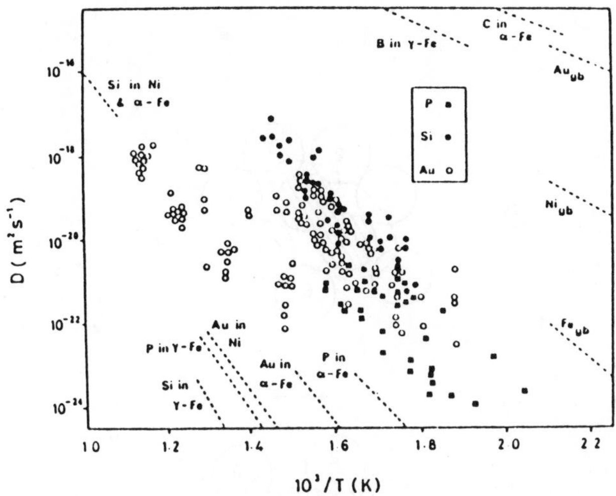

Fig.2. The diffusion coefficients for P, Si and Au in amorphous alloys, compared with the coefficients for: interstitial diffusion of C in α-Fe and B in γ-Fe; for grain boundary diffusion in Au, Ni and Fe; and for substitutional diffusion of a variety of solutes in α-Fe, γ-Fe and Ni. (After Cantor [16]).

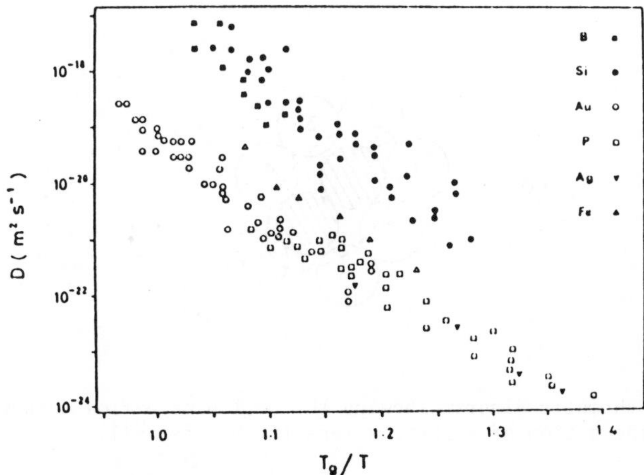

Fig.3. The diffusion coefficients of various elements in metal-metalloid amorphous alloys. The temperature has been normalised using the glass transition temperature T_g. (After Cantor [16]).

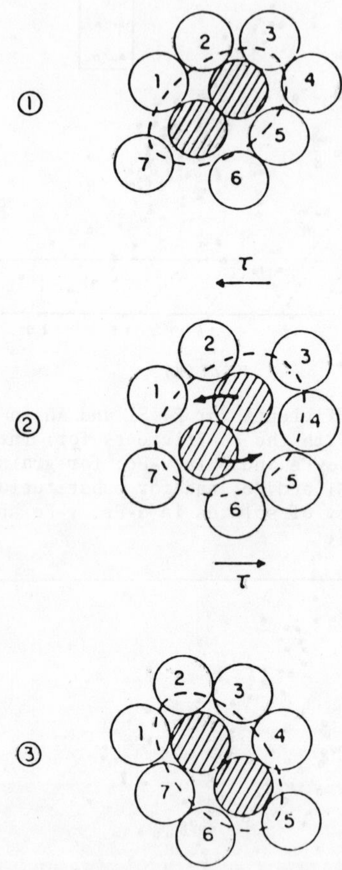

Fig.4. Schematic diagram showing the motion of a flow defect under the action of a shear stress (from Greer [9]).

A PHASE WITH FORBIDDEN SYMMETRY PRODUCED BY RAPID QUENCHING

A. L. Greer

University of Cambridge
Department of Metallurgy and Materials Science
Pembroke Street, Cambridge, CB2 3QZ, U.K.

1. INTRODUCTION

Shechtman et al.[1] reported the existence of a phase showing extended icosahedral symmetry (point group $m\bar{3}\bar{5}$) in melt-spun ribbons of composition Al with 10-14 at. % Mn, Fe or Cr. The grains of this phase, up to 2 μm across, gave 5-fold, 3-fold and 2-fold axes in electron diffraction patterns consistent with the symmetry. The orientation and nature of the patterns was independent of the selected area within a grain, showing that there is a long range orientational order in the grains. Five-fold rotation symmetry in an infinite lattice is forbidden, yet the sharp diffraction patterns show that the phase cannot be amorphous. This conflict can be resolved in two basic ways: the phase may be crystalline, but multiply-twinned [2,3]; or it may have a previously unrealised "quasicrystalline" structure. There has been much interest in these phases, and the number of relevant papers in the literature is now over 200, a number which is expanding rapidly as new investigations are carried out and as earlier mathematical work is seen to be significant.

At Cambridge, high resolution electron microscopy of melt-spun $Al_{84}Mn_{16}$ alloy has been carried out mainly by Drs K.M. Knowles, W.M. Stobbs and W.O. Saxton. This work has published in detail [4-6] and only the main results are reported here.

2. RESULTS

An alloy of composition $Al_{86}Mn_{14}$ was melt-spun on a brass wheel in air. The cooling rate is estimated to be 10^5 to $1o^6$ Ks^{-1}. After electropolishing it was examined mainly in the Cambridge 600 kV high resolution electron microscope. This instrument has an axial contrast transfer function which extends without zeros beyond 0.18 nm at 575 kV under optimum conditions [7].

The melt-spun ribbons consisted of 0.5 μm grains of a phase showing icosahedral symmetry embedded in an aluminium matrix. Figure 1 is a high resolution image of the icosahedral phase taken using the Cambridge high resolution electron microscope at 500 kV near Scherzer defocus with the electron beam parallel to a 5-fold axis. From this, and similar images with the same beam direction a number of points can be made [4].

i) There is no evidence of multiple twinning. Dark field images do show local variations in intensity, but there do not appear to be any twinning defects or centres of symmetry. While it is not possible to rule out the presence of a multiply-twinned phase in other samples similarly prepared [2], an explanation of icosahedral symmetry based on the twinning of a large-unit-cell structure [3] does not seem to be compatible with the HREM images.

ii) The structure (fig.1) is uniform. Though some features may be regarded as defects, there is no evidence for dislocations as postulated for icosahedral quasicrystals [8].

iii) The main reflections in the diffraction pattern correspond to a real space periodicity d = 0.206 nm. In the sets of parallel lines (in 5 orientations) evident in fig.1, spacings of τd and $\tau^2 d$ can be readily distinguished, where τ is the golden ratio $(\sqrt{5} + 1)/2$. These spacings can be regarded as short (S) and long (L) steps in the Fibonacci series generated by the rule S → L and L → LS. The sequences of steps in the micrograph obey the rules for such a series: the population of L steps is τ times that of S steps; forbidden sequences such as LLL and SS do not occur; and repeated deflation using the rules LS → L and L → S appears to give equally valid sequences.

iv) In thin areas (e.g. lower right in fig.1) the high resolution images consist of small rings with white dots at their centres. Interestingly, the size of these rings corresponds to that expected for the projection down the 5-fold axis of an icosahedral arrangement of 12 aluminium atoms around a central manganese.

v) When the resolution of the microscope is degraded by imposing high voltage ripple, very similar micrographs (both in thin areas, i.e. rings; and in thicker areas, i.e. Fibonacci series of lines) are obtained but with the features scaled up by a factor of τ. This shows that great care is needed before attempting direct interpretation of high resolution electron microscope images.

Uniform microstructures with no evidence of twinning were seen also in high resolution images with the electron beam parallel to symmetry axes other than the 5-fold [5]. Thus it seems clear that the icosahedral phase is not twinned, but has a special quasicrystalline structure. Many possible descriptions of such a structure have been proposed; a useful list is given by Mackay [9]. Of these the most attractive are those based on a three-dimensional Penrose tiling [10-13]. This pattern is composed of two "unit-cells": rhombohedra with the same edge length, one acute with $\alpha = 63.43^{o}$ (arc tan 2), the other obtuse with $\alpha = 116.57^{o}$. These rhombohedra fit together perfectly (without distortions or gaps) to fill space. The resulting structure has long range orientational order with icosahedral symmetry, but it is not periodic. Despite the lack of periodicity, the pattern does have long range positional order; in fact it is perfectly ordered, with zero configurational entropy. The Penrose packing appears to be closely related to the structure of the Al-Mn icosahedral phase. Projections of the packing down a 5-fold axis into two dimensions gives a tiling in which the same series of lines in the Fibonacci series may be seen as in fig.1. Furthermore, a simple decoration of the rhombohedra (one atom at each corner) gives diffraction patterns in which the reflection positions match (at least in the 5-fold and 3-fold patterns) those found in electron diffraction [13]. Obviously the intensities of the reflections will depend on the nature of the atomic decoration, and much effort has been expended in trying to elucidate this.

At Cambridge, Knowles et al. [5] showed that not only simulated high resolution images, but also simulated diffraction patterns were very sensitive to the atomic decoration. This is very promising in that calculated diffraction intensities should provide a critical test of structural models. However, in later work by Knowles and Stobbs [6] an exhaustive trial of different decorations failed to produce any satisfactory fit with experiment. The decorations were based mostly on edge- and face-centring of the rhombohedra. Given that Penrose tiling explains many features of the experimental results it still seems reasonable for it to be the basis of a description of the Al-Mn quasicrystalline structure. The decoration may be of two or more different types or there may be a statistical occupancy of the two types of rhombohedra. It may also be useful to consider the role of chemical inhomogeneity on a scale of 3 to 7 nm as is suggested by electron energy loss spectroscopy and by dark field imaging [6].

3. RELATIONSHIP BETWEEN ICOSAHEDRAL QUASICRYSTALLINE PHASES AND AMORPHOUS PHASES

Performing a calculation with Lennard-Jones potentials Frank [14] found that a cluster of thirteen atoms, consisting of a central atom with twelve neighbours at the apices of a regular icosahedron, has a lower internal energy than a cluster with either crystallographic close-packed structure. He suggested that such icosahedral clusters would be common in simple liquids and it has been usual since to suppose that this contributes to the resistance of glasses and liquids to crystal nucleation. The concept of non-crystallographic packings has been extended by Hoare and Pal [15] who have shown that larger clusters of tetrahedral or icosahedral symmetry, with a few tens of atoms, can be more stable than crystallographic counterparts. Hoare [16] has illustrated a number of icosahedral clusters, "amorphons" with several hundred atoms. These extend to dense packed systems the concepts of "vitrons" [17] and "amorphons" [18] developed for tetrahedrally bonded systems. Such structures have also been found in molecular dynamics and Monte Carlo simulations of cooled liquids [16]. These dense, stable, non-crystallographic clusters have been suggested as the basis of a description of the structure of amorphous alloys. The clusters have the common characteristic that they consist of packed tetrahedra. While tetrahedra are the densest possible packing units in three dimensions, they are not space-filling: even the nearest neighbour cage, twenty tetrahedra forming an icosahedron, has surface bond lengths 5% greater than those to the central atom. Clusters consisting of tetrahedra only cannot be indefinitely extended because of increasing strain. Any model for the structure of amorphous alloys based on an assembly of finite non-crystallographic clusters has the severe problems that the "connective tissue" between the clusters is not described and may lead to an overall density that is too low. On the other hand, some tendency to non-crystallographic (specifically, icosahedral) packing in liquids and amorphous alloys is highly plausible and may explain the resistance to crystal nucleation.

The "frustration" limiting the size of non-crystallographic clusters can be removed in curved space, in which tetrahedra can be space filling. Of particular interest is an ideal, icosahedral crystal ("polytope {3,3,5}") which can be constructed in the surface of a four-dimensional sphere. A metallic glass structure can be considered to arise when such a crystal is disrupted by a disordered array of disclinations [19-21]. The disclinations divide the structure into regions of short range {3,3,5} order and reduce the space curvature to zero. In fact, icosahedral atomic packings do occur also in crystals. They are found in the Frank-Kasper phases, the structure of which can be described as an ideal icosahedral crystal intersected by an ordered array of disclinations. Thus there may be a close link between metallic glasses and the Frank-Kasper phases. Nelson [20] has pointed out that since the -72° disclination lines of interest here cannot cross, a tangle of lines in a high temperature liquid will have difficulty ordering; a glass will result instead of the lower energy Frank-Kasper phase. The tangled disclination line model of metallic glass structure has been used with some success to fit experimental structure factors for amorphous, nominally one-component systems formed by vapour deposition (fig.2).

Quasicrystals, such as the Al-Mn phase of concern in this paper, give sharp Bragg diffraction maxima at reciprocal lattice vectors which can be expressed as integer linear combinations of twelve vectors of length K_o pointing to the vertices of an icosahedron. Among the most intense reflections are those at K_o, 1.052 K_o, 1.701 K_o and 2.0 K_o. If the sharp first peak found in the structure factor for an amorphous alloy is associated with the reflections at K_o and 1.052 K_o, the reflections at 1.701 K_o and 2.0 K_o correspond closely with the two components of the second peak (fig.2). Thus a quasicrystal, with extended icosahedral order (orientational and translational) and an amorphous alloy may have similar correlations of atomic position within their structures [22]. It has been pointed out, however, that the phase diagrams of alloy systems forming quasicrystalline and amorphous phases are quite different, and that close parallels between the two cases may be unfounded [23].

4. DISCUSSION

Icosahedral quasicrystalline phases have now been made by a number of methods other than melt-spinning (or in general rapid liquid quenching), e.g. by ion-implantation [24] and by devitrification of a metallic glass [25]. They have been found in many alloy systems, including some not based on aluminium. A reasonably comprehensive list is given in Table 1.

As shown in the table there are other phases related to the icosahedral quasicrystalline phase already described. The T-phase appears to be a periodic stacking of quasiperiodic planes [26]; it is often mistaken for the icosahedral phase. The τ-phase exhibits vacancy ordering and appears to be a quasiperiodic stacking of regular lattice planes [26]. Thus the three types of phase may be described respectively as being quasiperiodic in three, two and one dimensions.

The most striking feature of the icosahedral phase is the 5-fold symmetry axes, because these are forbidden in crystals. It is quite possible that other "forbidden" symmetries will be found, and already 12-fold symmetry has been claimed in a $Cr_{70.6}Ni_{29.4}$ alloy [27]. It is interesting to note that after the concept of quasicrystals with forbidden symmetries became established (by the experimental demonstration of Shechtman et al.[1]), rather than earlier theoretical analyses[12]),confirmation and further discoveries followed rapidly. Quasicrystalline phases must have been produced earlier, but, one may speculate, they must have been disregarded because the resultant diffraction patterns were regarded as impossible. There are some parallels here with the discovery by Duwez and coworkers in 1959 that rapid quenching of a liquid could produce an amorphous alloy[28]. Many amorphous alloys are now known, and one wonders whether the present modest list of known quasiperiodic alloys in Table 1 will grow similarly.

ACKNOWLEDGEMENTS

The electron microscopy studies would not have been possible without the alloy preparation by Mr J. Leader and the melt-spinning by Ms J.E. Rout. The author gratefully acknowledges collaboration with Drs K.M. Knowles, W.M. Stobbs and W.O. Saxton, and useful discussions with Dr J.A. Leake and with Professors D.R. Nelson and F. Spaepen. Thanks are also due to the S.E.R.C. for financial support and to Professor D. Hull for the provision of laboratory facilities.

REFERENCES

1) Shechtman, D., Blech, I., Gratias, D. and Cahn, J.W., Phys. Rev. Lett. 20, 1951 (1984).
2) Field, R.D. and Fraser, H.L., Mater. Sci. Engng. 68, L17 (1984/5).
3) Pauling, L., Nature 317, 512 (1985).
4) Knowles, K.M., Greer, A.L., Saxton, W.O. and Stobbs, W.M., Phil. Mag. B 52, L31 (1985).
5) Knowles, K.M., Saxton, W.O., Stobbs, W.M. and Greer, A.L., Inst. Phys. Conf. Ser. No. 78: ch 9, p.321 (1985).
6) Knowles, K.M. and Stobbs, W.M., Nature, submitted.
7) Smith, D.J., Camps, R.A., Freeman, L.A., Hill, R., Nixon, W.C. and Smith, K.C.A., J. Microscopy 130, 127 (1983).

8) Levine, D., Lubensky, T.C., Ostlund, S., Ramaswamy, S., Steinhardt, P.J. and Toner, J., Phys. Rev. Lett. 54, 1520 (1985).
9) Mackay, A.L. and Kramer, P., Nature 316, 17 (1985).
10) Penrose, R., Bull. Inst. Maths. Its Appl. 10, 266 (1974).
11) Mackay, A.L., Sov. Phys. Crystallogr. 26, 517 (1981).
12) Mackay, A.L., Physica 114A, 609 (1982).
13) Levine, D. and Steinhardt, P.J., Phys. Rev. Lett. 53, 2477 (1984).
14) Frank, F.C., Proc. Roy. Soc. A 215, 43 (1952).
15) Hoare, M.R. and Pal, P., Adv. Phys. 20, 161 (1971).
16) Hoare, M.R., Ann. N.Y. Acad. Sci. 279, 186 (1976).
17) Tilton, L.W., J. Res. Nat. Bur. Stand. 59, 139 (1957).
18) Coleman, M.V. and Thomas, D.J.D., Phys. Status Solidi 24, K111 (1967).
19) Sadoc, J.F., J. Phys. (Paris) Colloq. 41, C8, 326 (1980).
20) Nelson, D.R., Phys. Rev. B28, 5515 (1983).
21) Sethna, J.P., Phys. Rev. B31, 6275 (1985).
22) Nelson, D.R., in Proc. Workshop on Amorphous Metals and Semiconductors, San Diego, 1985 (eds. P. Haasen and R.I. Jaffee). Pergamon, Oxford, in press.
23) Phillips, J.C., J. Mater. Res. 1, 1 (1986).
24) Budai, J.D. and Aziz, M.J., Phys. Rev. B., 33, 2876 (1986).
25) Poon, S.J., Drehman, A.J. and Lawless, K.R., Phys. Rev. Lett. 55, 2324 (1985).
26) Chattopadhyay, K., Lele, S., Ranganathan, S., Subbana, G.N. and Thangaraj, N., Current Sci. 54, 895 (1985).
27) Ishimasa, T., Nissen, H.-U. and Fukano, Y., Phys. Rev. Lett. 55, 511 (1985).
28) Klement, K., Willens, R.H. and Duwez, P., Nature 187, 869 (1960).
29) Koskenmaki, D.C., Chen, H.S. and Rao, K.V., Phys. Rev. B33, 5328 (1986).
30) Rao, K.V., this conference.
31) Bancel, P.A., Heiney, P.A., Stephens, P.W., Goldman, A.I. and Horn, P.M., Phys. Rev. Lett. 54, 2422 (1985).
32) Ramachandrarao, P. and Sastry, G.V.S., Pramana 25, 225 (1985).

33) Sastry, G.V.S., Rao, V.V., Ramachandrarao, P. and Anantharaman, T. R., Scripta Metall. 20, 191 (1986).

34) Bendersky, L.A. and Biancaniello, R.S., unpublished.

35) Zhang, Z., Ye, H.Q. and Kuo, K.H., Phil. Mag. in press.

36) Shechtman, D., Schaefer, R.J. and Biancaniello, F.S., Metall.Trans. 75A, 1987 (1984).

37) Chattopadhyay, K., Ranganthan, S., Subbanna, G.N. and Thangaraj, N., Scripta Metall. 19, 767 (1985).

38) Sastry, G.V.S., Suryanarayana, C., van Sande, M. and van Tendeloo, G., Mater. Res. Bull. 13, 1064 (1978).

39) Sastry, G.V.S., Suryanarayana, C. and van Tendeloo, G., Phys. Stat. Solidi (a) 73, 267 (1982).

40) van Sande, M., de Ritter, R., van Landuyt, J. and Amelinckx, S., Phys. Stat. Solidi (a) 50, 587 (1978).

Table 1.

Compositions of phases exhibiting quasiperiodicity

Quasiperiodic in: 3-D		2-D		1-D	
Icosahedral phases		T-phases		τ-phases	
Al(lo-14) at.% Mn	1)	Al-Mn	26,36,37)	Al-Pd	39)
Al-(lo-14) at.% Fe	1)	Al-Pd	38)	Al-Cu-Ni	40)
Al-(10-14) at.% Cr	1)	Al-Mn-Ni	38)		
Al-Mn-Si	29)				
Al-V	30)				
Al-Pd	31)				
Al-Pt	31)				
Al-Ru	31)				
Al-Mn-Cr-Si					
Al-Mn-Ru-Si					
$Mg_{32}(zn,Al)_{49}$	32)				
$Pd_{60}U_{20}Si_{20}$	25)				
Mg_4CuAl_6	33)				
Cd-Cu	34)				
$(Ti_{1-x}V_x)_2Ni$	35)				

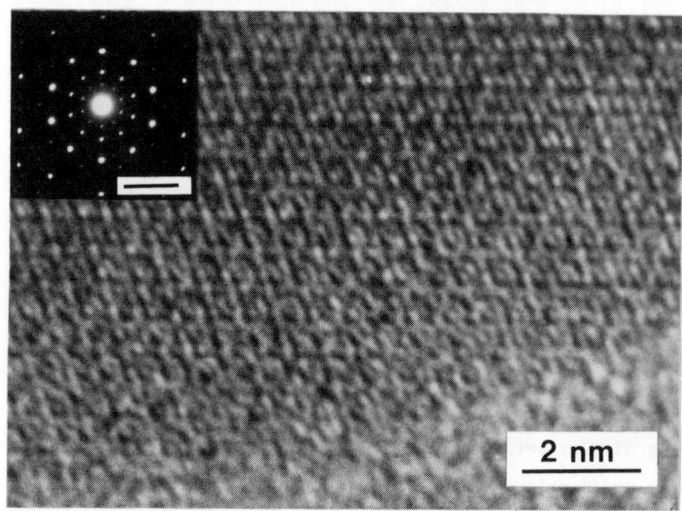

Fig.1. High resolution electron micrograph and electron diffraction pattern from the icosahedral ($m\bar{3}\bar{5}$) phase in a rapidly quenched $Al_{86}Mn_{14}$ alloy. The electron beam is parallel to a 5-fold axis. The scale bar on the diffraction pattern represents 5 nm^{-1}.

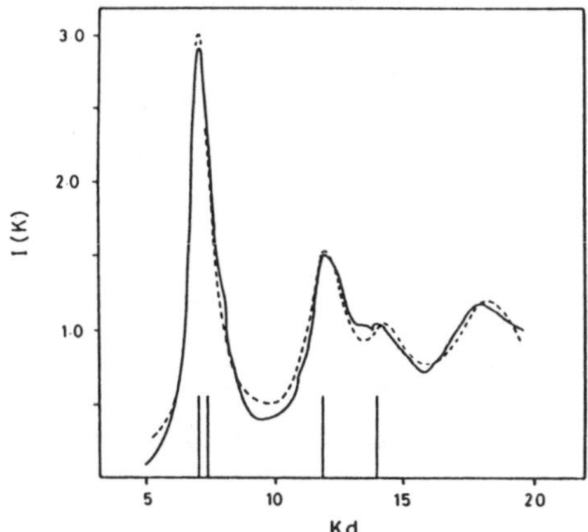

Fig.2. Structure factors for amorphous iron, as measured (solid line), and as calculated (dashed line) assuming an ideal icosahedral crystal (with interparticle spacing d) disrupted by a random array of disclinations. The four most prominent reflections from an icosahedral quasicrystal are also marked. (After [22]).

EFFECT OF COMPOSITION AND QUENCHING RATE ON STRUCTURAL RELAXATION OF METALLIC GLASSES

Giuseppe Riontino

Universita di Torino - Via P. Giuria, 9 - 101125 Torino (Italy)

In general the term "structural relaxation" implies the complex of phenomena which occur during annealing of an amorphous system at the temperatures lower than the glass-transition (as in some cases) or below the crystallization temperature. Structural relaxation modifies several physical properties of the material, mainly the mechanical and magnetic ones, namely the density, the hardness, the Young modulus, the coercivity, the permeability etc. Usually the variations of properties do not exceed 10% for a given composition; so, from technological point of view, the relevant value is often of the same order of magnitude. However, from scientific point of view, these variations are in effect much more interesting, since they involve interactions between the constituent atoms.

Starting from a metastable condition with respect to the corresponding crystalline state, the amorphous structure reduces on annealing its free energy undergoing small rearrangements through short or medium range ordering. These ordering is usually specified as the chemical ordering, when the atomic species rearrange themselves without appreciable variations in density, and the topological ordering, more collective in nature, by which the overall structure is compacted towards the final products of crystallization.

It is clear that the kinetics of structural relaxation depends on chemical composition of the alloy, since the chemical affinity of the constituent atoms as well as their diffusivity influence the motion of atoms or group of atoms during in thermal treatment.

The initial state of the amorphous is another important parameter to be taken into account since it is rate-determining for the successive, thermally activated processes leading to crystallization. This state depends on the preparative parameter. For metallic glasses obtained by rapid quenching from melt, the quenching rate is the most effective factor in determining the initial state of metastability for a given chemical composition of the alloy. To differentiate macroscopically the initial conditions, a measure can be made of the free volume frozen-in during the quench, i.e. the difference between the volume of the amorphous and of the liquid undercooled in thermodynamic equilibrium conditions. It can be explained microscopically, (Cohen and Grest [1)]), that an infinite liquid-like cluster exists in liquid; when it is undercooled, a fraction of liquid-like cells is still present and its amount depends on the quenching rate, being lower for the lower quenching rates. Recently Fujita [2)] has assumed that a medium range order is present in liquids; he calculated on the basis of the rate process theory the concentration of the frozen-in cluster in the quenched liquid, and he found it being dependent on the quenching rate. These are only some examples of modelling, just to show how much attention is paid to the starting structural conditions of an amorphous, which in turn, are the result of, chemical bonding of the constituent elements.

Different experimental techniques are used to follow the process of structural relaxation; I shall discuss here the variations in electrical resistivity during thermal treatments at temperatures higher than ambient. It is difficult to explain exactly, exclusively on this basis, what happens in an amorphous system: there does not exist yet a comprehensive theoretical frame for interpretation of these data. Notwithstanding the complexity of the phenomena leading to structural rearrangements in the process, the electrical resistivity may give a valuable contribution to the solution of the problem.

First of all, I will show the essential lines of a simple formalism connecting the free volume content after quenching and the electrical resistivity variations around the room temperature. Details on the subjects are reported in refs. [3,4].

The initial linear trend observed for the relative variation of $\Delta\rho/\rho$ as a function of linearly increasing temperature, in various Fe-based amorphous alloys (Figs. 1,2,3) can be justified if starting from an extension of the Ziman theory of electrical resistivity ρ in liquid alloys, with some assumptions on the structure factor for resistivity [5]. Calculating the temperature coefficient for resistivity (TCR) α, i.e., the initial slope of $\Delta\rho/\rho$ as a function of T around the room temperature, one derives firstly that α is positive, as in our case, when the actual value of the Fermi wave vector is not strictly coincident with the position of the first peak in the radial distribution function.

The change of α with the free volume content may be derived from the original approach of Srolovitz et al. [6], who explain the structural relaxation in terms of annihilation of local stress - density fluctuations present in as-quenched samples. In this model, regions where the density is higher or lower than the average one, are characterized by non-equilibrium values of the short-range hydrostatic pressure p, which is randomly distributed in the material with an average null value and a non-zero second moment $\langle p^2 \rangle$. The value of $\langle p^2 \rangle$ is found to be proportional to the macroscopic free volume and it is modified during structural relaxation [7]. The radial distribution function G(r) for atoms under hydrostatic pressure p, may be expressed in terms of $G_o(r)$, i.e. the RDF for atoms with p = 0; on this basis the structure factor in the expression for α is found to depend on the $\langle p^2 \rangle$ value.

Since the changes in $\langle p^2 \rangle$ (initially proposed to explain the free volume reduction during structural relaxation induced by annealing) may be related to the free volume content frozen-in on quenching, variations in α have to be observed also in the case of variable quenching rate. In effect, one obtains that the $\Delta\alpha$ is proportional to $\Delta\langle p^2 \rangle$, and the proportionality constant can be calculated using the Percus-Yevick model [8] of hard spheres packing, which suitably reproduces the feature of the first peak of the structure factor. One finds that the value of the constant is negative in the same range of the wave vectors for which $\Delta\langle p^2\rangle$ is positive. We can conclude that at negative values of $\Delta\langle p^2\rangle$ (i.e. at lower quenching rates) the higher values of $\Delta\alpha$ should be obtained. This can be seen from the results in Fig.4 for different types of glassy alloys, all with positive α. The average ribbon thicknesses, being inversely proportional to the quenching rate, are given.

It should be stressed that only qualitative evaluation of $\langle p^2\rangle$, and therefore of the absolute free volume content, may be extracted from the experimental α values. In this model, the physical parameters in the expression of ρ and α must be known for an exact determination of $\langle p^2\rangle$.

As regards the effect of composition, the electrical resistivity variations can also provide some useful indications about the mechanism of structural relaxation though only qualitative at this stage of research. The starting point is the different behaviour displayed by various types of glassy alloys at intermediate temperatures of annealing, between room temperature and crystallization point. It can bee seen in Fig.1, presenting the results for a series of amorphous FeNiMoB alloys with different Mo content, that some deviations from linearity of $\Delta\rho/\rho$ at intermediate temperatures are evidenced, proved not to be due to magnetic transformations (as the occurrence of the Curie temperature). The same trend is displayed by the amorphous FeNiCrPB series with variable Cr content. The results are presented in Fig.2. Finally the $\Delta\rho/\rho$ is shown in Fig.3 for FeB-based glassy alloys, in which a small amount of transition metals is substituted for iron.

At first these examples will be considered. The bendings are more pronounced in the case of Cr or Mo containing alloys with respect to the corresponding master alloys. The same is for FeB alloys containing Ti, V, Cr or Mn with respect to those containing Co or Ni, and the master alloys itself. Different mechanisms seem to be involved in thermal ageing at intermediate temperatures, where structural relaxation occurs, thus all samples were treated isothermally in order to be able to distinguish different contributions to the relaxation. The hypothesis can be made, in fact, that the presence of particular atomic species and their interactions with the amorphizing elements (like P or B) can modify the rearrangements of the basic amorphous structure. Some typical results of the resistivity variations with respect to its value at the beginning of the isotherm, as a function of annealing time are shown for the Mo series in Fig.5, for the Cr series in Fig.6 and for the transition metals series in Fig.7. A decrease as well as an increase is observed depending on the annealing temperature and on the element characterizing the series.

The resistivity decrease could be explained by the commonly accepted reduction of free volume due to the topological short range ordering during structural relaxation. The resistivity decrease at times (or temperatures) at which the crystallization begins is a normal general feature of all samples. For the oposite behaviour, the hypothesis can be made that Cr and Mo, on one hand, and the elements on the left from iron in the periodic table, on the other hand, favour this kind of chemical short range ordering which has effects similar to those observed during pre-precipitation in crystalline alloys, i.e. an increase of the electrical resistivity with time. Of course, concurrently free volume reduction and densification of structure may occur, which should lead to decrease in resistivity as in the above case; but chemical clustering, having an opposite effect, causes an increase of the resultant resistivity. This behaviour is observed at low temperatures, when not so high energies are required to activate the small displacements involved in chemical ordering. At higher temperatures more collective processes of free volume reduction take place, and the prevailing effects of resistivity decrease may be observed. At intermediate temperatures, the two effects can balance themselves, and practically no variation is evidenced during the annealing time.

Since we have no direct experimental evidence of such clustering during relaxation, I have to call "inductive" the considerations of this kind. For the amorphous series containing transition metals, SAXS and Mössbauer measurements have been performed [9)], and the results suggest that structural inhomogeneities present in as-quenched as well as in the annealed samples may account for the resistivity behaviour during isothermal annealing. In effect, it is shown that in the case of

Fe, Co and Ni-containing alloys a partial chemical order, already present in the as-quenched samples, favours topological rearrangements not explicitly involving the ordering of particular atomic species, thus accounting for the observed resistivity decrease at any annealing temperature. For the other alloys, the compositional ordering, not significantly affecting the topology of the amorphous configuration, is believed to prevail at the first stage of annealing. This can be due to the reduced initial mobility of the whole structure, and it is also in agreement with the higher activation energies required to crystallize these alloys [9)]. One can conclude that chemical ordering seems to be a preliminary condition for the reorganization of the amorphous structure. Its degree of competitivity with topological ordering depends on chemical affinities between the constituent atoms.

An alternative approach to the problem has been represented recently by Kelton and Spaepen [10)] to explain the isothermal behaviour of the resistivity variations. They combined the Ziman theory (confirmed to be a good approximation for a wide range of initial resistivity values), and their own free volume model of the viscous flow. They derived an expression for $\Delta R/R$ as a function of annealing time; some parameters in this expression can be obtained by fitting the resistivity data. These parameters can be compared with the results on the viscosity relaxation, and good agreement is found for various amorphous alloys. They conclude that the linear time dependence of the viscosity relaxation can account for time dependence of the resistivity relaxation, also in the case in which an increase of the latter has been observed. The assumption is made that the changes of these two properties are different manifestations of the same structural changes.

Apart from some minor discrepancies revealed by the Authors on the kinetic parameters obtained in fitting the data (in effect explained by themselves), one can remark that the model suggests that the increased resistivity, when it occurs, is related to the corresponding value of the TCR α, which has to be negative in these cases, as effectively found.

In our case, in which α is always positive, it seems that other factors besides the free volume reduction have to be considered during structural relaxation. At present there are several indications that chemical clustering is common in glassy alloys obtained by rapid quenching from melt. So it is not surprissing that these cluster act as the scattering centres for the conductivity electrons, until the thermal motions make the amorphous structure generally more "ordered".

REFERENCES

1) Cohen, M.H. and Grest, G.S., Phys. Rev. B20, 1077 (1979).
2) Fujita, F.E., Proc. V Int.Conf. on Rapidly Quenched Metals, eds. S. Steeb & H. Warlimont, Elsevier Sci. Pub. B.V., vol. 1, 585, 1985.
3) Allia, P., Andreone, D., Sato Turtelli, R., Vinai, F. and Riontino, G., J. Appl. Phys. 53, 8798 (1982).
4) Allia, P., Sato Turtelli, R., Vinai, F. and Riontino, G., Solid State Comm., 43, 821 (1982).
5) Meisel, L.V. and Cote, P.J., Phys. Rev. B17, 4652 (1978).
6) Srolovitz, D., Maeda, K., Vitek, V. and Egami, T., Philos. Mag. 44, 847 (1981).
7) Srolovitz, D., Egami, T. and Vitek, V., Philos. Rev. B24, 6936 (1981).
8) Cote, P.J. and Meisel, L.V., Topics in Applied Physics, 40 (Glassy Metals I), Springer Verlag 155 (1981) and references therein.
9) Matteazzi, P., Cocco, G., Le Caer, G. and Riontino, G., in "Industrial Applications of the Mössbauer Effect", eds. J. Stevens & G.J. Long., Plenum Press Co., to appear.
10) Kelton, K.F. Spaepen, F., Phys. Rev. B30, 5516 (1984).

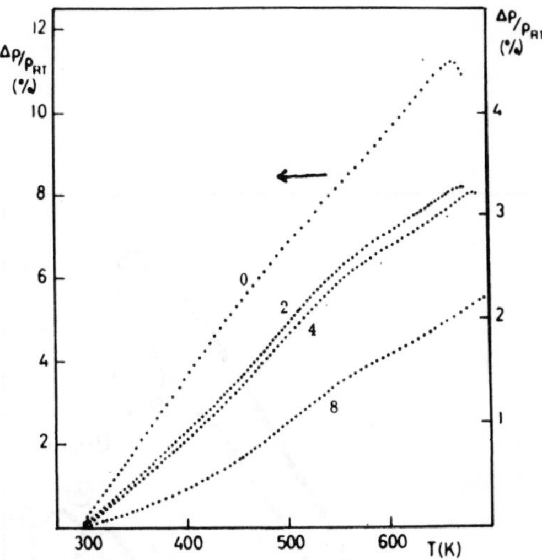

Fig.1. Relative variations of resistivity with respect to its value at room temperature, as a function of linearly increasing temperature, for $Fe_{63-x}Ni_{23}Mo_xB_{14}$ amorphous alloys; x = 0,2, 4,8 at %, respectively.

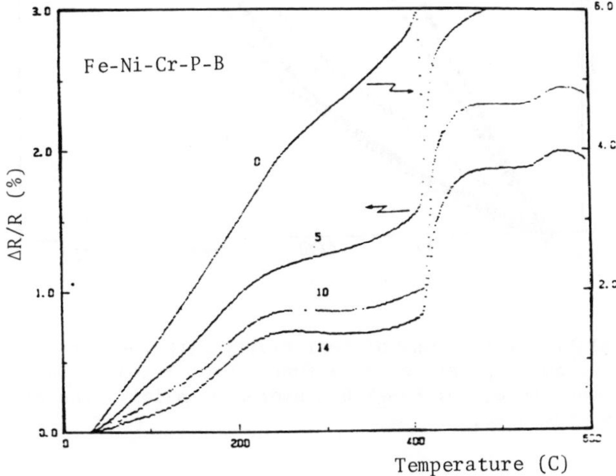

Fig.2. Relative variations of resistivity with respect to its value at room temperature, as a function of linearly increasing temperature for $Fe_{44-x}Ni_{36}Cr_xP_{14}B_6$ glassy alloys; x = 0,5, 10 and 14 at %, respectively.

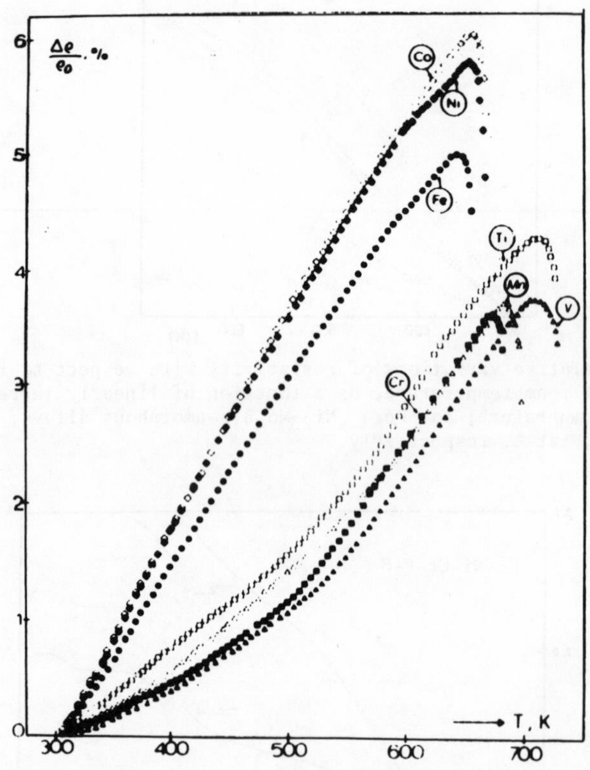

Fig.3. Relative variations of resistivity with respect to its value at room temperature, as a function of linearly increasing temperature, for $Fe_{75}X_5B_{20}$ amorphous alloys. The element X is indicated.

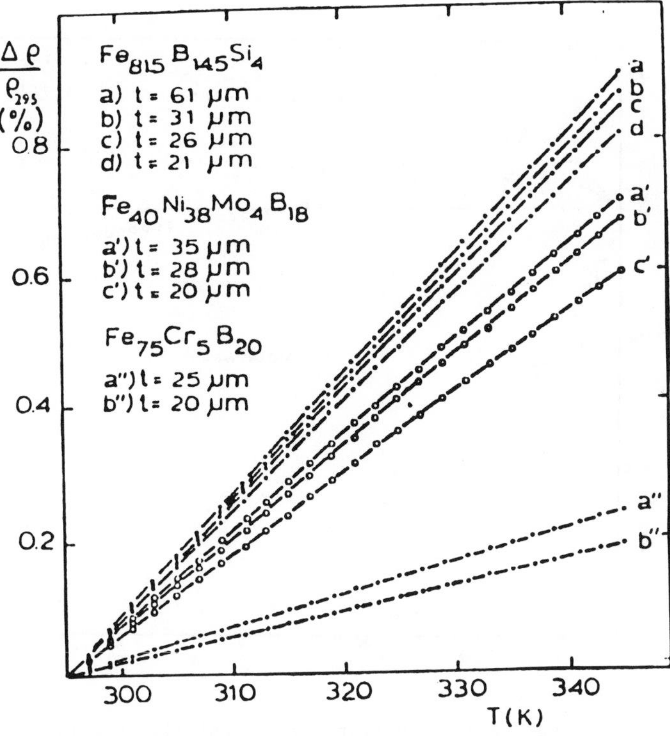

Fig.4. Relative variations of resistivity with respect to its room temperature value, as a function of temperature. The amorphous compositions and the ribbon thicknesses are indicated.

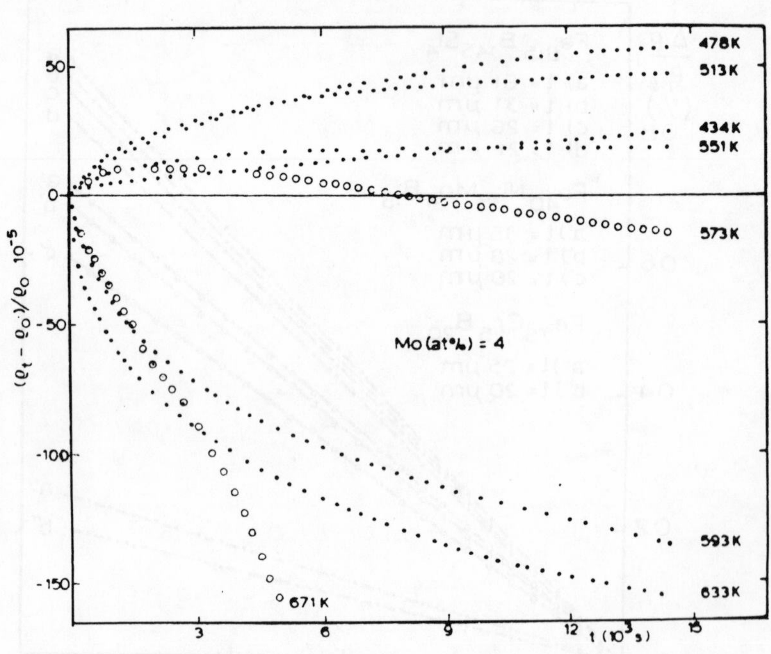

Fig.5. Typical relative variations of resistivity with respect to its initial value as a function of annealing time, for the Mo-containing alloys.

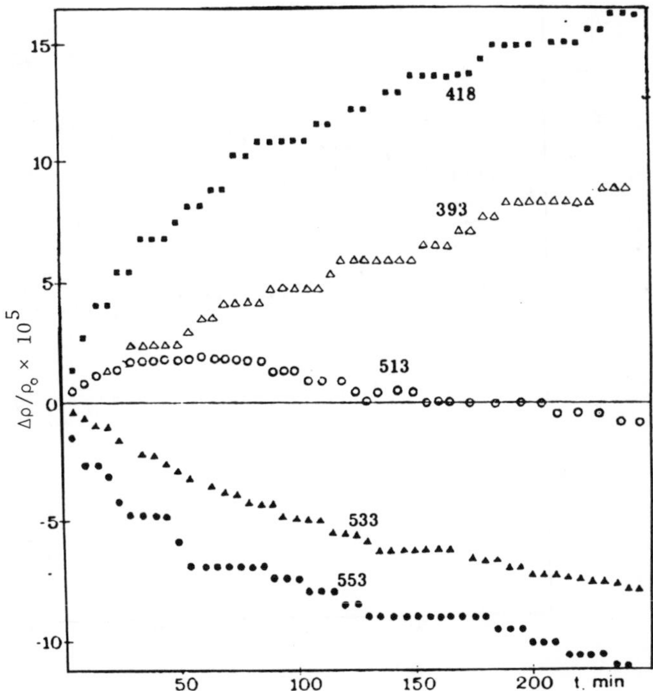

Fig.6. Typical relative variations of resistivity with respect to its initial value, as a function of annealing time for Cr-containing alloys.

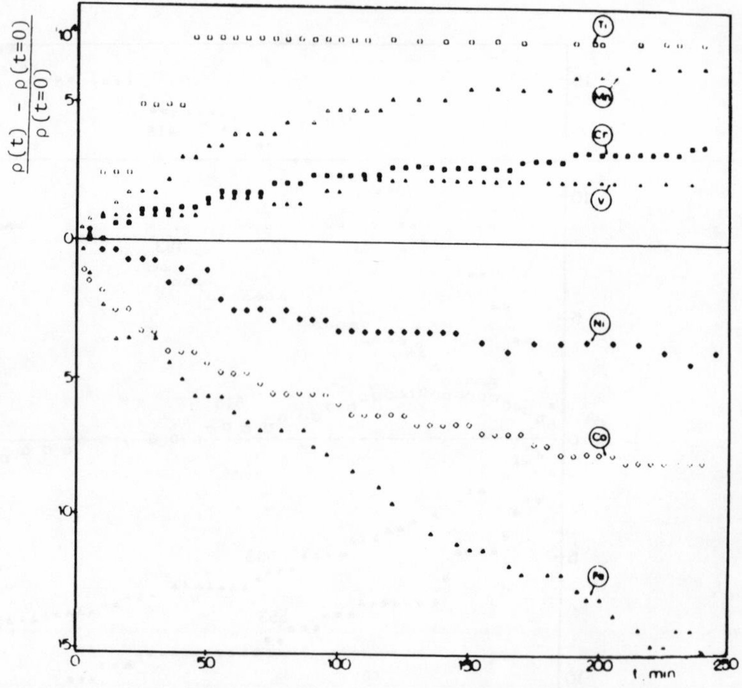

Fig.7. Typical relative variations of resistivity with respect to its initial value, as a function of annealing time for transition metal-containing alloys. The substituting elements are indicated.

DENSITY OF METALLIC GLASSES

G. Konczos and B. Sas

Central Research Institute for Physics, Hungarian Academy of Sciences, 1525 Budapest 114 POB 49, Hungary

1. INTRODUCTION

Density and molar volume are some of basic physical properties of solids. This is especially true in the case of the metallic glasses where no long-range order is preserved. Usually the validity of a given structural model is first checked by the density data. Density is frequently used to determine the volume of irregularly shaped small samples.

Experimental data on the density of amorphous alloys (mainly the metal-metalloid glasses) are available, but not much information can be found on the critical interpretation of density and the derivative quantities, e.g. the packing fraction and the mean atomic volume [1-3].

This short review deals only with two aspects of density of metallic glasses. The experimental technique for measuring the density of small samples is briefly discussed and some typical examples are mentioned concerning the use of the density data.

2. MEASUREMENT OF THE DENSITY OF METALLIC GLASSES

There is a great variety of techniques for measuring the density of solids, including the frequently used Archimedes method. In this method the weight of a specimen is determined in two different media, e.g. in air and liquid. Weighing in air is a simple procedure and can relatively easily be performed with the use of a modern electronic microbalance. Weighing in liquid is much more complex and this limits to great extent the reproducibility of the measurements. Bowman et al. [4] found that the major loss of precision is due to nonreproducibility of the surface tension forces (meniscus effect) and to the trapping of gas bubbles on the surface of the specimen.

The precision is also dependent upon the mass of the sample. The density of samples with large masses can be determined with a high degree of accuracy. Bowman et al. [4] were able to obtain a reproducibility of 1 ppm in the density measurements of silicon single crystals with a mass of about 20 grams. Metallic glasses are generally produced in the form of thin ribbons or wires and these forms are not really adequate for the measurement of density. In particular, it is frequently necessary to measure the density of small samples (30-50 mg) [5]. In such cases a reproducibility of 0.3-0.6% can be achieved in a simple set-up. This precision is usually sufficient for a number of practical purposes, e.g. the determination of the volume of sample for magnetic measurements, the study of the dependence of density on composition, etc.

In this connection a simple technique has been developed by the authors to measure the density of metallic glasses. A block diagram of the apparatus is presented in Fig.1. The ME3 microbalance (METTLER Co., Switzerland) provides an accuracy of about ± 1 μg. Using this balance the sample can be suspended outside the balance, under the table, on a fine gold-plated tungsten wire (50 μm dia.) connected with the balance with a quartz fibre. Toluene is used as an immersion liquid. The meniscus effect is determined in a separate experiment and is taken into account as a correction. A typical correction value, while using gold plated tungsten wire of 50 μm diameter, is of about 0,420 mg.

Organic liquids of higher densities are frequently used by different authors (e.g. bromoform or tetrabromethane of densities of about 2.0 and 2.5 g/ml, respectively). According to experience of the present authors the meniscus of these liquids is not so stable as that of toluene, moreover the bromoform slowly decomposes, even at room temperature. Therefore their use in density measurements has some disadvantages.

We have achieved a precision of about 0.5% in the case of wide metallic glass ribbons weighing 50-100 mg. Cawthorne and Sinclair [6] developed a more sophisticated method giving an accuracy of about ±0.05% for 30 mg samples. The experimental methods for the determination of density on small samples are discussed in details by Pratten[5].

3. RESULTS

3.1 Dependence of Density on Composition

The dependence of density upon composition has been determined in a great number of amorphous alloys, mainly in metal-metalloid glasses, such as binary Ni-P, Fe-P, Co-P, Fe-B, Co-B, ternary Fe-TM-B (TM: transition metal), Fe-Si-B, Co-Si-B and the multicomponent Co-based alloys (Table 1). The density is usually a linear function of the metalloid concentration. Anomalous behaviour has been found in the Fe-B system [7].

In structural studies not the density itself but some derivative quantities are used, namely the molar volume (V in ml/mol), the mean atomic volume and the packing fraction (η). Physical interpretation of these quantities is not very easy because we have no unambiguous procedure to distinguish the contributions representing the atomic volumes of the transition metal and the non-metal atoms.

For metal-metalloid glasses it is suggested [3] to divide the total volume between the metal atoms alone and then determine the metal atom packing fraction (η_M). η_M is generally a linear function of the composition, whose gradient varies with the nature of the metalloid; the limiting value ($x \to 0$) is almost constant and equal to 0.70-0.72.

Systematic data on the density of metal-metal glasses are scarce but most of them show a high packing fraction, e.g. $\eta = 0.728$ for different Zr-M glasses (M = Fe, Ni, Co, Cu) [26].

3.2. Correlation Between Density and Other Physical Properties

The density data when combined with some other physical quantities often provide important information as can be illustrated by the following examples.

Ray et al. found that the compositional dependence of the saturation magnetic moment μ_s is consistent with that of the density of iron-boron metallic glasses and reflects electron charge transfer from boron to iron [28]. Konczos et al. [14] studied the influence of tungsten and chromium on the density and mean magnetic moment in ternary Fe-W-B and Fe-Cr-B metallic glasses. The crystallization temperature, activation energies and densities were determined as a function of the composition by Dong et al. [29] on Ni-Zr glasses. The results could not be explained by a simple dense random packing of hard spheres model. Zedler and Lehmann found that the mass density and microhardness changes due to isochronal heat treatment are well correlated in Fe-B-Si-C metallic glass [30].

3.3 Change of Density Due To Various Treatments

The structure of amorphous alloys can be modified by several procedures, e.g. by heat treatments (crystallization or relaxation), by neutron irradiation, by cold working, etc.

A number of authors found that the densities of amorphous and crystalline phases are similar. The density difference is usually less than 2%, the crystalline phase being more dense. Thus a conclusion can be drawn that the local structure of both phases is similar. The data for many systems are available, e.g. for Fe-Si-B, Co-Si-B [30].

The determination of densities in the differently relaxed states is much more difficult because the change is estimated to be of about 0.1-0.2% compared to the as-quenched state. In spite of the experimental difficulties there are some recent data in the literature. Gerling and Wagner [31] found a density change of about 0.08% after heat treatment far below the crystallization temperature for Fe-Ni-B metallic glasses. Similar measurements on the same system after neutron irradiation were made by Toloui et al. [32]. They found that the density decreased after irradiation due to the formation of the excess free volume.

REFERENCES

1) Cargill III, G.S., Structure of Metallic Alloy Glasses in: Ehrenreich, H., Seitz, F and Turnbull D. (Eds.): Solid State Physics, vol. 30, 227 (1975) Academic Press, New York.
2) Turnbull, D., Scripta Metall. 11, 1131 (1977).
3) Gaskell, P.H., Acta Metall. 29, 1203 (1981).
4) Bowman, H.A., Schnoonover, R.M., Jones, M.W., J. Res. NBS 71C, 179 (1967).
5) Pratten, N.A., J. Mater. Sci. 16, 1737 (1981).
6) Cawthorne, C., Sinclair, W.D.J., J. Phys. E 5, 531 (1972).
7) Shirakawa, K., Waseda, Y., Masumoto, T.; Sci. Rep. RITU A 29, 229 (1981).
8) Mehra, M., Subbaro, E.C., Kapur, P.C., Hasegawa, R., IEEE Trans. Magn. MAG-16, 1141 (1980).
9) Ray, R., Hasegawa, R., Chou, C.-P., Davis, L.A., Scripta Metall. 11, 973 (1977).
10) Logan, J., Phys. Stat. Solidi (a) 32, 361 (1975).
11) Schmidt, T., Varga, L., Keméy, T., Konczos, G., Tompa, K., Kajcsos, Zs., Nucl. Instrum. Meth. 199, 359 (1982).
12) Chen, H.S., Waseda, Y., Aust, K.T., phys. stat. sol. (a) 65, 695 (1981).
13) Waseda, Y, Masumoto, T., ibid. 31, 477 (1975).
14) Konczos, G., Kisdi-Koszó, É., Lovas, A., Kajcsos, Zs., Potocky, L., Daniel-Szabó, J., Kovac, J., Novak, L., J. Magn. Magn. Mat. 41, 122 (1984).
15) Waseda, Y., Aust, K.T., Walter, J.L. J. Mater. Sci. 15, 1252 (1980).
16) O'Handley, R.C., Hasegawa, R., Ray, R., Chou, C.-P, Appl. Phys. Lett. 29, 330 (1976).
17) Johnson, W.L., Williams, A.R., Phys. Rev. B 20, 1640 (1979).

18) Baczewski, T., Lipinsky, E., Appl. Phys. A 30, 213 (1983).
19) Aso, K., Hayakawa, M., Hotai, K., Vedaira, S., Makino, Y., Proc. 4th Int. Conf. on Rapidly Quenched Metals, eds. Masumoto, T. and Suzuki, K., p.379 (Japan Institute of Metals, Sendai 1982).
20) Luborsky, F.E., Frischmann, P.G., Johnson, L.A., J. Magn. Magn. Mat. 19, 130 (1980).
21) Goto, M., Tange, H., Tokunaga, T., Jpn. J. Appl. Phys. 18, 2023 (1979).
22) Kohmoto, O., Ohya, K., J. Appl. Phys. 57, 626 (1985).
23) Hasegawa, R., Tanner, L.E., Phys. Rev. B 16, 3925 (1977).
24) Altounian, Z., Strom-Olsen, J.O., Phys. Rev. B 27, 4149 (1983).
25) Calvayrac, Y., Chevalier, J.-P., Harmelin, M., Quivy, A., Bigot, J. Phil. Mag. B 48, 323 (1983).
26) Krebs, H.U., Freyhardt, H.C., Proc. 5th Int. Conf. on Rapidly Quenched Metals, eds. Steeb, S., Warlimont, H, p.439 (Nort-Holland, Amsterdam, 1985).
27) Calvayrac, Y., Quivy, A., Chevalier, J.-P., Bigot, J., ibid. p.455.
28) Hasegawa, R., Ray, R., J. Appl. Phys. 49, 4174 (1978).
29) Dong, Y.D., Gregan, G., Scott M.G., J. Non-Cryst. Solids, 43, 403 (1981).
30) Zedler, E., Lehmann, G., Proc. 5th Int. Conf. on Rapidly Quenched Metals, eds. S. Steeb and H. Warlimont, p. 743 (North-Holland, Amsterdam, 1985].
31) Gerling, R., Wagner, R., Scripta Metall. 16, 963 (1982).
32) Toloui, B., Kursumovic, A., Cahn, R.W., Scripta Metall. 19, 947 (1985).

Table 1.

References on densities of some metallic glasses

Metallic glasses	References
Co-B	7
Fe-B	7 - 9
Co-P	1, 7
Fe-P	7, 10
Ni-P	1, 11
Pt-P	12
Pd-Si	13
Fe-Cr-B	14
Fe-Ni-B	15,16
Fe-W-B	14
Mo-Ru-B	17
Fe-P-B	7
Co-Si-B	7, 18, 19
Fe-Si-B	20
Cu-Pd-P	17
Mn-Pd-P	1
Ni-Pd-P	1
Ni-Pt-P	1
Cu-Pd-Si	1
Mo-Ru-Si	17
Fe-Co-Si-B	7, 21, 19, 22
Fe-Ni-Si-B	7, 21
Ni-Co-Si-B	7, 19
Be-Zr	23
Co-Zr	24
Cu-Zr	24, 25
Fe-Zr	24, 26
Ni-Zr	24
Cu-Ni-Zr	27

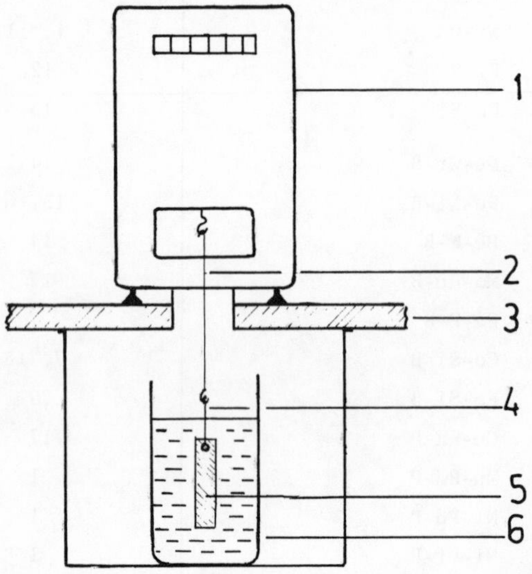

Fig.1. Simple apparatus for density measurement of small samples: 1 - digital electronic microbalance, 2 - quartz fibre, 3 - table, 4 - gold plated tungsten wire, 5 - sample, 6 - glass filled with organic liquid.

CRYSTALLISATION OF METALLIC GLASSES

Uwe Köster

Department of Chemical Engineering, University
Dortmund D-4600 Dortmund 50, F.R.Germany

1. INTRODUCTION

Metallic glasses have attracted a lot of attention in recent years, since they exhibit a combination of remarkable properties, which are directly derived from their glassy state [1], e.g. soft ferromagnetism, relatively high electrical resistivity, very high tensile strength and excellent corrosion behaviour. However, it is becoming increasingly clear, that similar to glass ceramics some partially crystallised metallic glasses possess even improved properties, e.g. higher ductility [2], less magnetic core loss for high frequency applications[3], or stronger flux pinning interactions in superconducting metallic glasses [4]. Fully crystallised metallic glasses specially designed for the purpose can exhibit an unique microstructure consisting of ultrafine grains (<1 μm), stabilized by extremely finely dispersed borides, with impressive high-temperature properties or even superplasticity [5,6]. Recently, new hard magnetic materials with unusually high energy products were produced by crystallisation of melt-spun praseodymium- and neodymium-iron-boron glass precursors [7].

During the last years crystallisation of metallic glasses has become a subject of an increasing research effort and has been reviewed by a number of authors [8,...,11]. Impeding or controlling crystallisation, both are of utmost importance for any application of metallic glasses, and crystallisation studies are of interest:

- in understanding glass formation and assessing the thermal stability of metallic glasses,
- in elucidating the fundamentals of nucleation and growth processes during crystallisation,
- in learning to produce particular novel and useful microstructures by controlled crystallisation unobtainable by other means.

2. THERMODYNAMICS OF CRYSTALLISATION:

Metallic glasses crystallise by nucleation and growth processes. The driving force is the difference in free energy between the glass and the appropriate crystalline phase(s). In order to gain an overall picture of all possible crystallisation reactions a (hypothetical) diagram of the free energy for the various phases versus concentration has been found to be very useful [12]. Fig. 1 shows such a diagram for the Fe-B system; depending on the composition crystallisation can occur by:

- polymorphic crystallisation of one phase with the same composition as the glass. This reaction can occur only in concentration ranges near the pure elements or compounds and proceeds by single jumps of atoms across the crystallisation front.
- primary crystallisation of one crystal phase with a composition different from the glassy matrix, e.g. α-iron. During this reaction a concentration gradient is built up ahead of the crystallisation front and the amorphous phase will be enriched in boron until crystallisation is stopped after reaching the metastable equilibrium; subsequently, the boron enriched matrix transforms by one of the other mechanisms discussed here, e.g. polymorphic crystallisation.
- eutectic crystallisation, where two crystalline phases grow cooperatively by a discontinuous reaction. There is no difference in the overall concentration across the reaction front; diffusion takes place parallel to the reaction front and the two components have to separate into the two phases, thus reducing the growth rate of this reaction compared with that during polymorphic crystallisation.

These three reactions are shown schematically in fig. 2.
Appropriate free energy vs. composition diagrams can be constructed from available thermodynamical data of the melt and the crystalline phases by assuming the amorphous alloy as an extension of the liquid phase to temperatures in the undercooled region [13]. If not enough thermodynamical data are available, regular solution behaviour might be assumed and the heat of mixing of the liquid phase can be estimated using the method of Miedema [14,15]. The difference in free energy ΔG between the amorphous and the appropriate crystalline phase can be calculated from the observed enthalpy of fusion ΔH_f and the undercooling ΔT; an approximation for ΔG is given by the following equation [16]:

$$\Delta G = \frac{\Delta H_f \cdot \Delta T}{T_f} \cdot \left(\frac{2T}{T_f + T}\right) \tag{1}$$

Due to the driving force (difference in free energy) most metallic glasses can crystallise by two or even more different reactions. Which of those crystallisation reactions occurs, depends not only on the driving force but also on the kinetics of each. In most cases the equilibrium phases do not form directly from the glass but crystallisation proceeds by the formation of a sequence of metastable phases.

3. CRYSTAL GROWTH

Crystal growth may be primary, eutectic or polymorphic and has been observed to be always thermally activated. Growth rates can be estimated from crystal diameter distributions after appropriate annealing times; in-situ observations are misleading due to changes in the reaction mode and the kinetics at or near surfaces. Depending on the crystallisation mode we can distinguish between parabolic and linear growth [9]:

3.1 Parabolic Growth

Primary crystallisation is generally assumed to be controlled by volume diffusion. Assuming a concentration independent diffusion rate D the radius r of a spherical particle will be proportional with the $\sqrt{time}$, i.e.:

$$r = \alpha \cdot \sqrt{D \cdot t} \tag{2a}$$

and the growth rate u_p will be given by:

$$u_p = \frac{1}{2} \cdot \alpha^2 \cdot D \cdot r^{-1} \tag{2b}$$

where α is a dimensionless parameter evaluated from the composition at the particle interface and the composition of the sample. This diameter-dependent growth rate will be reduced drastically as soon as the diffusion fields of adjacent particles overlap and the regime of Ostwald ripening is achieved.

On the other hand, this relation can be used to calculate diffusivities from primary crystallisation data as shown in fig. 3 for $Fe_{40}Ni_{40}B_{20}$ glasses, where primary crystallisation [17] is controlled by boron diffusion; the iron diffusion measured by means of the radiotracer technique in combination with ion-beam sputtering [18] is about one order of magnitude slower. In a number of glasses the temperature dependence of the diffusivity D has been found to be equal to:

$$D = D_o \cdot \exp(- \frac{Q_D}{R \cdot T}) \tag{3}$$

Prefactors D_o and activation energies for diffusion Q_D are listed in table 1.

3.2 Linear Growth:

Linear growth, i.e. constant growth rates, have been found to occur in crystallisation reactions which do not involve any compositional change (polymorphic or partitionless crystallisation) or in a co-operative transformation into two crystalline phases involving long-range diffusion in such a way that the mean composition of the crystallised region is equal to that of the glassy matrix (eutectic crystallisation).

Once a stable nucleus has formed, it grows during isothermal annealing at large undercooling with a rate u_1 given by:

$$u_1 = u_o \cdot \exp(-\frac{Q_g}{R \cdot T}) \qquad (4)$$

where Q_g is the activation energy for growth. The preexponential u_o equals to:

$$u_o = \delta \cdot \nu \qquad (5)$$

where ν is a characteristic frequency in the order of the Debye-frequency and δ the distance across the crystallisation front. With an interface thickness δ of about 0.5 nm the prefactor will be in the order of $5 \cdot 10^3$ m/s.

For a number of metallic glasses prefactors u_o as well as the activation energies for growth during eutectic and polymorphic crystallisation are shown in table 2; the observed preexponentials u_o are normally orders of magnitude larger than predicted by the above theory, especially in ternary or more complicated glasses.

Activation energies and prefactors for eutectic crystallisation are only effective ones, since the growth rate depends not only on the temperature, but also on the interlamellar spacing λ which itself is a function of the temperature [8)]; the observed increase in interlamellar spacing with decreasing temperature (compare fig. 4a with 4b) can be understood due to increasing difficulties in nucleation of the second phase.

But even for polymorphic reactions prefactors and activation energies are by orders of magnitude too large. Extremely large pre-exponentials have been observed to be always correlated with very high activation energies and might therefore indicate growth by simultaneous transfer of a group of atoms from the glassy to the crystalline phase rather than singly. Other explanations assume that impurities are present and the temperature dependence of the growth rate does not only give the activation enthalpy for boundary movement.

The large prefactors have been found to occur only in systems with anisotropic crystal growth. Such a behaviour, however, indicate a growth mechanism by the lateral motion of steps; if the fraction of these active sites is assumed to be x_a, the growth rate in equation (4) has simply to be multiplied by x_a. If we assume that the number of these active sites x_a increases with the annealing temperature, we are able to explain an effective higher activation energy and a larger prefactor[9].

4. NUCLEATION

Nucleation phenomena can be investigated only indirectly, e.g. by analysing the microstructure after some amount of crystallisation by extrapolating growth reactions to shorter times [21]. Complementary techniques such as DSC (differential scanning calorimetry) provide data on reactions kinetics in a larger sample volume.
Nucleation for primary and polymorphic crystallisation are single step reactions, but eutectic crystallisation has been observed to be a process with two nucleation steps starting in transition metal-boron glasses usually with the formation of the boride. The analysis of primary crystallisation is complicated due to the mentioned radius dependence of the growth rate. Therefore, we will concentrate in the following part on nucleation for polymorphic crystallisation.
In classic nucleation theory the steady state homogeneous nucleation rate I_{st} is given by:

$$I_{st} = I_o \cdot \exp(-\frac{L \cdot \Delta G_c}{R \cdot T}) \cdot \exp(-\frac{Q_N}{R \cdot T}) \qquad (6)$$

where I_o is a constant factor, L the Loschmidt number, Q_N the activation energy for the transfer of atoms across the liquid/nucleus interface, and ΔG_c the free energy required to form a nucleus of critical size. For heterogeneous nucleation at a limited number N_o of nucleation sites the activation energy ΔG_c is reduced due to the gain in surface energy thus leading to a significant higher nucleation rate I^*_{st}.
At the very beginning of annealing a finite period is expected during which the steady-state distribution of clusters assumed in classical nucleation theory is established. This involves a transient or time-de-

pendent nucleation rate I(t) which can be calculated from the following equation from Kashiev [22]:

$$I(t) = I_{st}\cdot\{1+2\sum_{n=1}^{\infty}(-1)^n\cdot\exp(-n^2\cdot\frac{t}{\tau})\} \tag{7}$$

where τ is the time-lag, which is expected to increase significantly with decreasing temperature.

Nucleation can be studied in some detail by crystallisation statistics of partially crystallised glasses, i.e. from the fit of calculated and measured crystal diameter distributions [23,24]: Let us assume polymorphic crystallisation of spherical crystals, a constant growth rate u, and apply the following calculations only for the initial states of crystallisation, when non-interfering crystals are formed. Then, depending on the operating nucleation mode, the number of crystals ΔN_i nucleated during the time-interval Δt is given by the following equations (see also fig. 5):

- Pre-existing post-critical nuclei: If there is a limited number Z of pre-existing post-critical nuclei, they will start growing as soon as the annealing temperature is reached. This will lead to a very narrow diameter distribution.
- Homogeneous nucleation with a rate I_{st} in steady state:

$$\Delta N_i = I_{st}\cdot(1-X_{i-1})\cdot\Delta t$$

$$\text{with:}\quad X_i = \frac{4\pi}{3}\cdot u^3\cdot\sum_{j=1}^{i}\Delta N_j\cdot\{\Delta t\cdot(i+1-j)\}^3 \tag{8}$$

- Heterogeneous nucleation at a limited number N_o of active nucleation sites with a rate $\overset{*}{I}_{st}$ in steady state:

$$\Delta N_i = \overset{*}{I}_{st}\cdot(1-X_{i-1})\cdot\{1-N_o^{-1}\cdot\sum_{j=1}^{i}\Delta N_j\}\cdot\Delta t$$

$$\text{for } i:\quad \sum_{j=1}^{i}\Delta N_j < N_o;\ \Delta N_i = 0 \text{ for all other } i. \tag{9}$$

The distribution in fig. 5c is typical for heterogeneous nucleation

with a rate in steady state; for a transient type heterogeneous nucleation $\dot{I}^*_{st}$ in equation (9) is replaced by the time-dependent rate $\dot{I}^*(t)$ and this distribution changes into that shown in fig. 5e.
In $Fe_{65}Ni_{10}B_{25}$ glasses [24)] polymorphic crystallisation of prolate $(Fe,Ni)_3B$-crystals has been observed as shown in fig. 6a. The size distributions of these crystals after isothermal annealing were determined using the method of van't Hoff [25)] and is plotted for one annealing treatment in fig. 7. Such a distribution is typical for a heterogeneous nucleation process with transient rates. The agreement with the diameter distributions calculated by crystallisation statistics is very good. From such fits of experimental measured and calculated crystal diameter distributions the time-lag τ and the steady-state nucleation rate $\dot{I}^*_{st}$ can be determined for different annealing temperatures (see fig. 8a).

The mode of nucleation observed during polymorphic crystallisation of metal-metalloid-glasses has been found to depend strongly on the annealing temperature [24)]:
Below T_g we observe a transient-type heterogeneous nucleation. The number of crystals per unit volume does not change with the annealing temperature; the temperature dependence of I_{st} and τ is of Arrhenius type. The activation energies (Q_N = 300 kJ/mole; Q_τ = 283 kJ/mole) are in reasonable accordance with the activation energy for diffusion in similar (Fe,Ni)-B glasses (Q_D = 260 kJ/mole [17)]. Whereas the number of quenched-in nucleation sites decreases strongly with increasing cooling rate during the casting process, both, time-lag τ as well as steady state nucleation rate $\dot{I}^*_{st}$ have been found to be independent.
Above the glass transition temperature T_g the number of crystals increases by orders of magnitude (see fig. 6b) probably due to homogeneous nucleation [26)]. Even in this temperature range transient effects with a significant time-lag τ have been observed. The large crystal in fig. 6b, which is assumed to be formed by heterogeneous nucleation in the very beginning of annealing, can be used for calculating the time-lag τ for the homogeneous nucleation mode: in $Fe_{65}Ni_{10}B_{25}$ glasses τ ranges between 4 s at 480°C and 10^{-2} s at 550°C.

The observed temperature dependence for the nucleation rate in this temperature range above the glass transition temperature T_g indicates an activation energy for nucleation of more than 1000 kJ/mole; such a high activation energy is far from understood. There exists, however, a striking similarity between the temperature dependence of the nucleation rate and that of viscosity (see fig. 8) [24], which may indicate that nucleation above T_g is controlled by viscous flow rather than diffusion of single atoms. If one assumes such a behaviour, diffusivity in the equation (6) for the nucleation rate has to be replaced by viscosity whose temperature dependence is given usually be the Fulcher-Vogel equation:

$$\eta = A\cdot\exp\left(\frac{B}{T-T_o}\right) \tag{10}$$

and the temperature dependence of I_{st} should be given by:

$$I_{st} \propto \exp\left(-\frac{L\cdot\Delta G_c}{R\cdot T}\right)\cdot\exp\left(\frac{-B}{T-T_o}\right) \tag{11}$$

In the temperature range between 460 and 530°C, where $L\cdot\Delta G_c$ is assumed to be still small compared to $R\cdot T$, the parameter B and T_o can be estimated to be 4.700 K and 631 K, respectively, which is close to the values predicted for (Fe,Ni)-B glasses by Davies [27].

5. SURFACE INDUCED CRYSTALLISATION

Understanding surface crystallisation is of utmost importance for a number of applications of metallic glasses, e.g., corrosion resistance or catalysis; surface crystals may act as pinning center and increase magnetic eddy current losses in high-frequency application [3]. Magnetic anisotropy after annealing is assumed to originate from a compressive stress due to the formation of crystalline surface layers with a higher density [28]. Designing microstructures by controlled crystallisation may require suppression of surface induced crystallisation.

Surfaces are expected to catalize nucleation as the crystalline phase replaces a portion of the surface, thus reducing the total surface energy required for nucleation. The stress energy due to the volume change during crystallisation may be reduced at surfaces by a reduction in the thickness of the crystalline phase. The oxygen content near the surface may stabilize a number of crystalline phases, thus increasing the driving force for crystallisation. Selective oxidation of one of the components, however, may possess the strongest influence on surface crystallisation of metallic glasses [29].

Selective oxidation of the metalloids at the surfaces of metal-metalloid glasses is assumed to be responsible for the excessive crystallisation of e.g. α-iron [29]. Even at temperatures far below any crystallisation event in the bulk glass at the surfaces primary crystallisation of the transition metal has been observed in a number of iron, nickel or cobalt based metal-metalloid glasses with metalloid contents up to about 25 at.%. Chromium additions to the $Fe_{79}B_{16}Si_5$ glass have been reported to lead to the formation of a protective layer of chromium oxide at the ribbon surface during annealing, thus reducing the oxidation of the metalloids and therefore the preferred α-iron crystallisation at the surface [30].

$Fe_{39}Ni_{39}B_{22}$ metallic glasses (fig. 9) exhibit preferred crystallisation at the contact side along the grooves, i.e. areas which had the best contact to the quenching wheel and therefore the highest cooling rates [26,31): Removing both surface layers, which are assumed to contain a large number of quenched-in nuclei or nucleation sites, by electrolytical polishing lead even to enhanced crystallisation because of the change of composition induced by selective oxidation of iron during the removal process (see fig. 10); ion beam milling instead of the electrolytical polishing does not lead to such anomalies. Coating of the ion beam cleaned surface with a thin vapour deposited iron layer has been found to reduce the influence of preferred oxidation, whereas a nickel layer accelerates again surface nucleation.

REFERENCES

1) Cahn, R.W., Contemp. Phys. 21, 43 (1980)

2) Hillenbrand, H.-G., Hornbogen, E., Köster, U., Metall 36, 1059 (1982)

3) Datta, A., DeCristofaro, N.J., Davis, L.A., Proc. Rapidly Quenched Metals 4, 1007 (Sendai 1981)

4) Johnson, W.L., in: "Glassy Metals I", ed. Güntherodt, H.-J, Beck, H., Topics in Applied Physics, Vol. 46, 191, (Springer-Verlag 1981)

5) Ray, R., J. Mat. Sci. 6 (1981), 2924 & 2927

6) Ashdown, C.P., Zhang, Y.-C., Grant, N.J., Proc. Rapidly Quenched Metals 5, 1763 (Würzburg 1984)

7) Koon, N.C., Das, B.N., J. Appl. Phys. 55, 2063 (1984)

8) Köster, U., Herold, U., in: "Glassy Metals I", ed. Güntherodt, H.-J., Beck, H., Topics in Applied Physics, Vol. 46, 225 (Springer-Verlag 1981)

9) Köster, U., in: "Phase Transformations in Crystalline and Amorphous Alloys", Proc. Disc. Meeting Clausthal 1982, ed. Mordike, B.L., 113, (DGM, Oberursel 1983)

10) Scott, M., in: "Amorphous Metallic Alloys", ed. Luborsky, F.E., 144 (Butterworths, London 1983)

11) Köster, U., Metallkde, Z., 75, 691 (1984)

12) Herold, U., Koster, U., Proc. Rapidly Quenched Metals 3, 281 (Brighton 1978)

13) Köster, U., Weiss, P., J.Non-Cryst. Sol. 17, 359 (1975)

14) Schwarz, R.B., Wong, K.L. Johnson, W.L., J.Non-Cryst. Solids 61 & 62, 129 (1984)

15) Miedema, A.R., Philips Technical Review 36, 217 (1976)

16) Thompson, C.V., Spaepen, F., Acta Met. 27, 1855 (1979)

17) Köster, U., Herold, U., Becker, A., Proc. Rapidly Quenched Metals 4, 587 (Sendai 1981)

18) Horvath, J., Pfahler, K., Ulfert, W., Frank, W., Mehrer, H., Proc. NCM-3 (Grenoble 1985), in press

19) Kijek, M., Ahmadzadeh, M., Cantor, B., Cahn, R.W., Scripta Met. 14, 1337 (1980)

20) Brodowski, H., Sagunski, H., Z. Phys. Chem. NF 139, 149 (1984)

21) Köster U., Herold, U., Proc. Rapidly Quenched Metals 4, 717 (Sendai 1981)

22) Gutzow, I., Toschev, S., in: "Advances in Nucleation and Crystallization of Glasses", ed. Hench, L.L., 10, 1971

23) Köster. U., Blanke, H., Scripta Met. 17, 495 (1983)
24) Blanke, H., Koster, U., Proc. Rapidly Quenched Metals 5, 277 (Würzburg 1984)
25) Underwood, E.E., "Quantitative Stereology", Addison-Wesley Publ. Co., 1970
26) Herold, U., Dr.-thesis, Dept. Mech. Eng., Ruhr-Universität Bochum 1982
27) Davies, H., Proc. Rapidly Quenched Metals 3, 1 (Brighton 1978)
28) Ok, H.N., Morrish, A.H., Phys. Rev. B23, 1835 (1981)
29) Köster, U., Kristall & Technik 14, 1369 (1979)
30) Nathasingh, D.M., J. Appl. Phys. 55, 1793 (1984)
31) Köster, U., Herold, U., Hillenbrand, H.-G., Scripta Met. 17, 867 (1983)

Table 1: Diffusion data estimated from primary crystallisation

primary crystallisation	$Co_{80}B_{20}$	$Fe_{86}B_{14}$	$Fe_{84}C_8B_8$	$Fe_{90}Zr_{10}$	$Fe_{42}Ni_{42}B_{16}$
D_o $\|m^2/s\|$	$4\cdot10^{-3}$	$2\cdot10^{-4}$	$1.1\cdot10^{-3}$	20	1100
Q_D $\|kJ/mole\|$	193	180	180	305	260

Table 2: Crystallisation parameter for polymorphic and eutectic crystallisation of metallic glasses

polymorphic crystallisation	$Fe_{76}B_{24}$	$Fe_{65}Ni_{10}B_{25}$	$Ni_{66}B_{34}$	$Fe_{30}Zr_{70}$	$Co_{33}Zr_{67}$
u_o $\|m/s\|$	$7\cdot10^{6}$	$9\cdot10^{13}$	$1.1\cdot10^{8}$	$5\cdot10^{3}$	$4\cdot10^{5}$
Q_g $\|kJ/mole\|$	200	300	205	175	205
N_v $\|m^{-3}\|$	$5\cdot10^{17}$	$8\cdot10^{16}$	$7\cdot10^{14}$	homogeneous nucleation	homogeneous nucleation

eutectic crystallisation	$Fe_{80}B_{20}$	$Fe_{40}Ni_{40}B_{20}$	$Fe_{78}Mo_2B_{20}$	$Fe_{80}P_{14}B_6$	$Co_{70}B_{30}$
u_o $\|m/s\|$	$6\cdot10^{5}$	$5\cdot10^{11}$	$1,3\cdot10^{10}$	$4\cdot10^{11}$	$9\cdot10^{8}$
Q_g $\|kJ/mole\|$	200	253	270	275	220
N_V $\|m^{-3}\|$	$5\cdot10^{17}$	$\sim 10^{13}$	$1\cdot10^{16}$	$1\cdot10^{16}$	$1\cdot10^{15}$

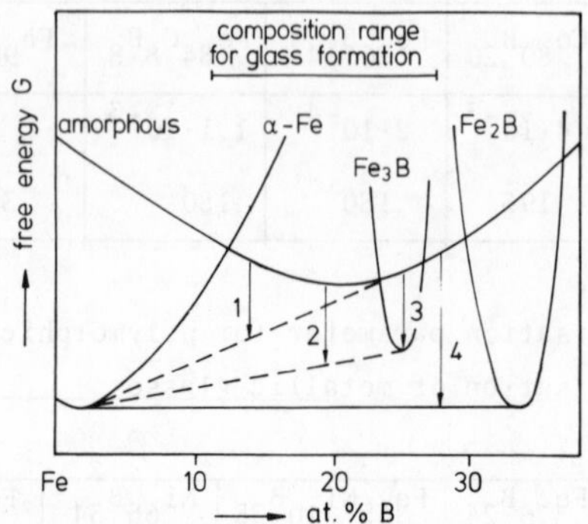

Fig. 1: Hypothetical diagram of the free energy for the various phases in Fe-B alloys versus concentration; the following crystallisation reactions can occur due to the difference in free energy [12]:
1) primary crystallisation of α-Fe;
2) eutectic crystallisation of α-Fe+Fe_3B;
3) polymorphic crystallisation of Fe_3B;
4) eutectic crystallisation of α-Fe+Fe_2B.

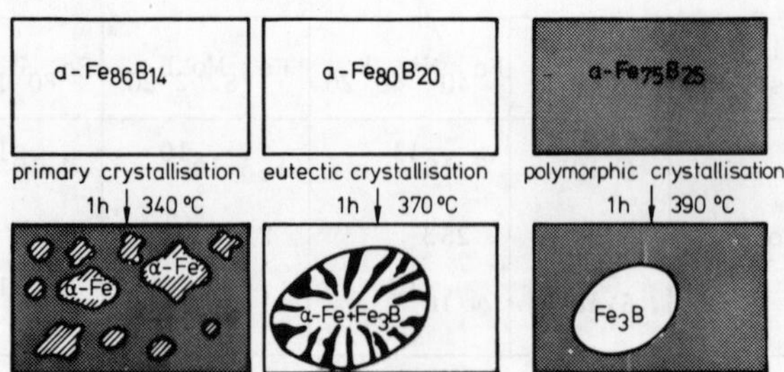

Fig. 2: Schematic diagram of typical crystallisation reactions in Fe-B metallic glasses [9].

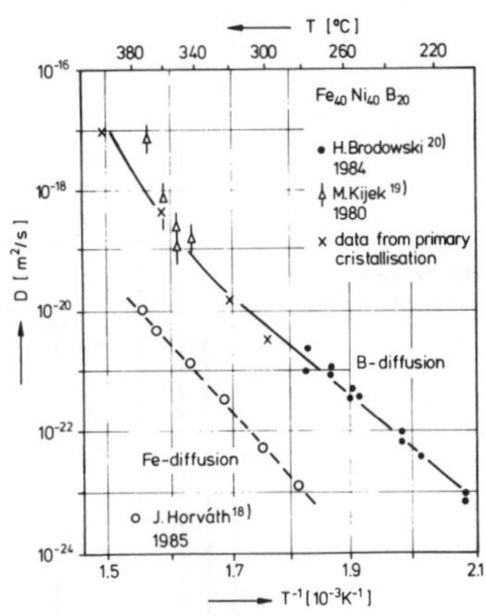

Fig. 3: Diffusion in relaxed $Fe_{40}Ni_{40}B_{20}$ glasses

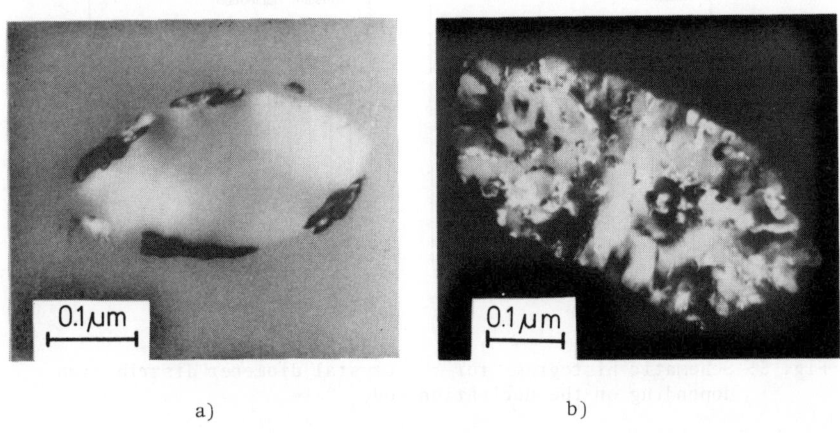

a) b)

Fig. 4: Early state of eutectic crystallisation in $Fe_{70}Ni_{10}B_{20}$ glasses: α-Fe crystallises at a $(Fe,Ni)_3B$-crystal.
a) 184 h at 310°C; b) 1 h at 380°C

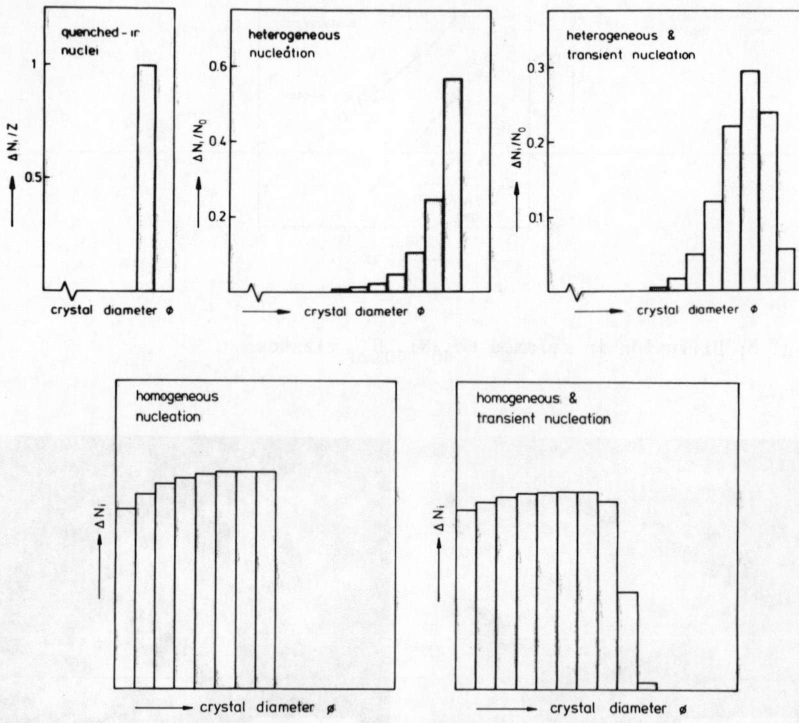

Fig. 5: Schematic histograms for the crystal diameter distribution depending on the nucleation mode [24]

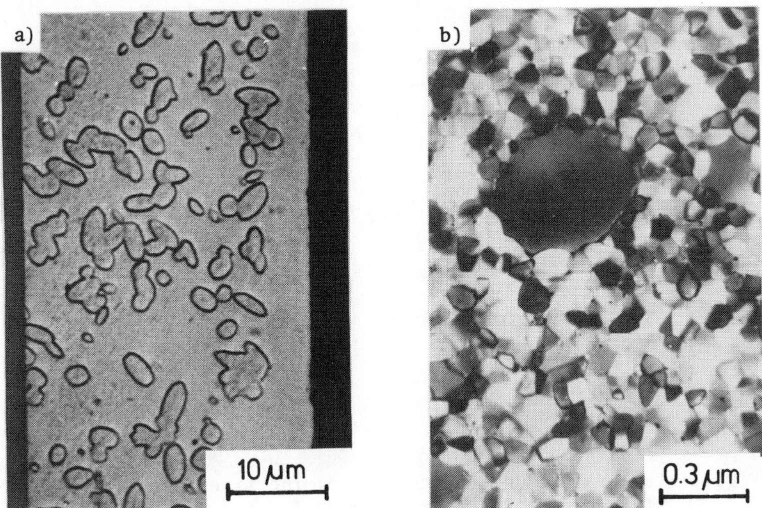

Fig. 6: Polymorphic crystallisation in $Fe_{65}Ni_{10}B_{25}$ glasses: a) below the glass transition temperature T_g: 350 min at 360°C (light micrograph/cross section); b) above the glass transition temperature: 10 s at 550°C (transmission electron micrograph)

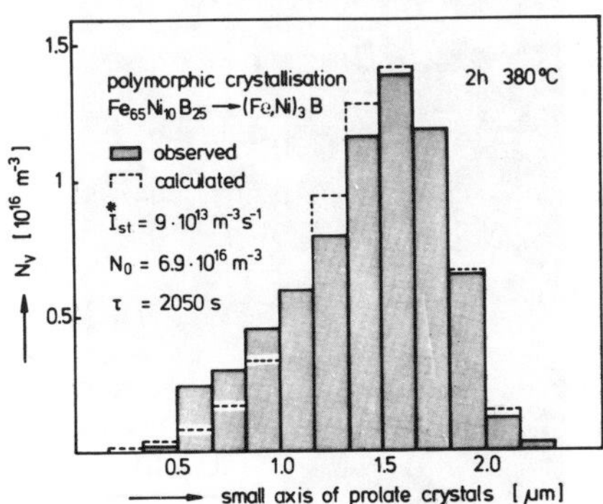

Fig. 7: Crystal diameter distribution (small axis) in partially crystallised $Fe_{65}Ni_{10}B_{25}$ glasses [24]

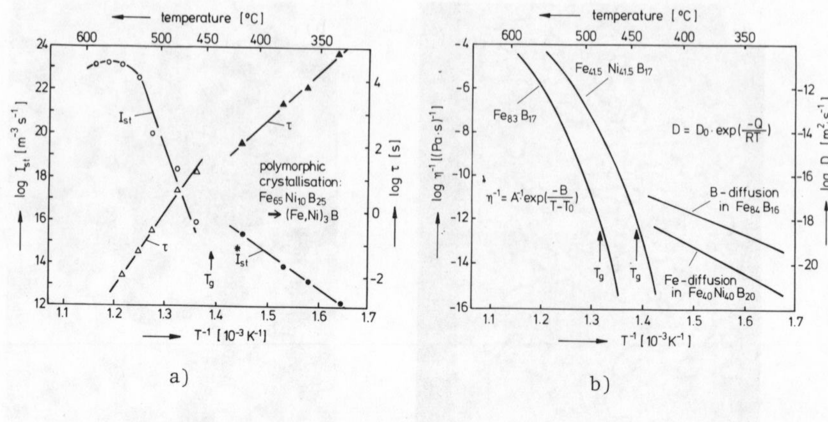

Fig. 8: Temperature dependence for a) the time-lag τ and the steady-state nucleation rates I_{st} and I^*_{st} and b) the viscosity η and the diffusion in Fe-Ni-B glasses 24)

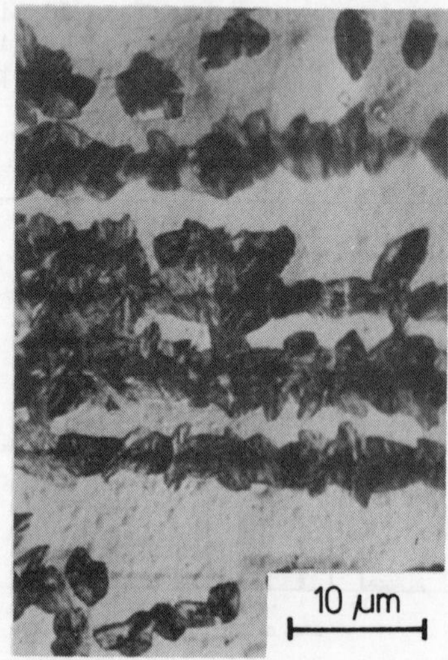

Fig. 9: Surface induced crystallisation at the contact side of a melt-spun $Fe_{39}Ni_{39}B_{22}$ ribbon: 20 min at 380°C

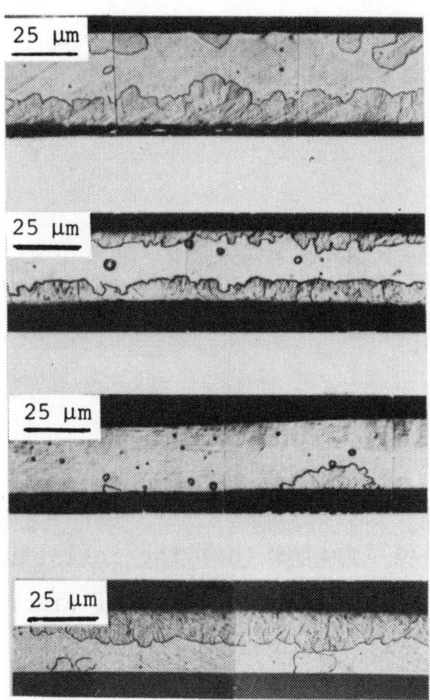

Fig. 10: Influence of surface treatments on the surface induced crystallization of melt-spun $Fe_{39}Ni_{39}B_{22}$ ribbons [26] (70 min at 390°C): a) as cast ribbon; b) annealing after electropolishing; c) annealing after ion beam milling, and d) annealing after ion beam milling followed by Ni deposition onto the top (free) surface of the ribbon.

THE ELECTRON TRANSPORT PROPERTIES OF AMORPHOUS METALS

J.O. Strom-Olsen

McGill University, Ernest Rutherford Physics Building,
3600 University Street, Montreal, Quebec, Canada, H3A 2T8

1. INTRODUCTION

According to the most elementary of all models of metals, the Drude-Sommerfeld-Lorenz model, the electrical resistivity of a metal [1] is given by

$$\rho = \sigma^{-1} = \frac{ne^2 V_F \ell^{-1}}{m} \qquad (1)$$

where ℓ the m.f.p. is a few interionic distances, say 5 Å. Since $V_F \sim 10^8 cm^{-1}$, $n \sim 10^{23} cm^{-3}$, this implies that the resistivity of a metal should be a temperature independent quantity of order 500 μΩ-cm. The thermoelectric power is given by the Mott formula [2]

$$S = \frac{\pi^2}{3} \frac{k_B^2 T}{|e|} \left. \frac{d\ln\sigma(E)}{dE} \right|_{E = E_F} \qquad (2)$$

which, using the same parameters as above and a few simple assumptions about the energy dependence of the quantities in r, gives s linear with temperature and of magnitude a few micro-volts per °K at room temperature. If we compare these predictions with a simple crystalline metal - say Al - we see at once that they are not even close: is a very strong function of T and is orders of magnitude smaller even at room temperature; S, although about the right size, is a very non-linear function of T (figure 1). If you now take a typical amorphous metal alloy - say Zr-Ni - its behavior is remarkably close to what is predicted by this model (figure 1) namely ρ is roughly constant and of order 200 $\mu\Omega$-cm, S is roughly linear with T and of order 3μV/K at 300K.

To zero order therefore amorphous metals may be viewed as simple metals in the Drude-Lorenz-Sommerfeld sense, and this simplicity derives directly from their lack of long range atomic order. Those who need nothing more than a rough idea of the magnitude of ρ and S need read no further because the rest of the article is concerned with how amorphous metals differ from these elementary rules.

The article is divided into four sections:

Resistivity and thermopower at high temperatures (>50K)

Hall effect

Resistivity and thermopower at low temperatures (<10K)

Some special effects in magnetic alloys.

The first three sections will concentrate almost exclusively on nonmagnetic metallic alloys because magnetic order can add extra effects over and above those arising from the disordered structure.

2. HIGH TEMPERATURE RESISTIVITY & THERMOPOWER

The general behavior of the resistivity of amorphous metals is similar to that in liquid metals. Table 1 shows the resistivity of a range of amorphous metals, together with other pertinent data. Several points may be made. The first is that, roughly speaking, the resistivity of amorphous metals lies in the decade above very impure crystalline metals and in the decade below more heavily doped semiconductors. Another feature is that the highest resistivities come from amorphous metals where the Fermi energy lies within a d-band. These metals (which we will now call d-band amorphous metals) in fact constitute the vast majority of amorphous metals. Within this group of d-band alloys there is a rather striking correlation between the resistivity and the valence electron density obtained by including <u>all</u> the valence electrons from unfilled shells, d as well as s. This correlation implies that to obtain an amorphous metal with high resistivity, one should choose a system where the d-band is unfilled and the electron density is low. A further correlation, better known, is that of the temperature coefficient of resistivity (TCR) with the magnitude of resistivity - known as the Mooij correlation [3] (figure 2), after the first observation of this phenomenon in liquid metals. In its original form the correlation stated that if the resistivity was greater than 150μΩ-cm then the TCR, $\frac{1}{\rho}\frac{d\rho}{dT} < 0$. Actually, as has been pointed out by Mizutani [4], there are really two Mooij correlations: one for simple amorphous alloys (with no d-band) (where the sign change occurs at about 50 μΩ-cm) and one for d-band amorphous alloys (where the sign change occurs at 150 μΩ-cm). In addition to the Mooij correlation, it is interesting to note specific variations of ρ with T. Figure 3 shows data for one simple (amorphous alloy) and two d-band alloys. The resistivity of simple amorphous metals varies as T^2 at temperatures well below the Debye temperature (but above the lowest temperature, see next section) giving way to T at higher temperatures; d-band amorphous metals generally have a quite different behavior and cannot be uniquely classified (the Cu-Ti alloy illustrated in figure 3 has been fitted to an expression of the form $\{A\text{-}B\ \exp(\frac{-T}{\Delta})\}$).

These are the salient features of the high temperature resistivity of amorphous metals. What do theories predict? The simplest and most developed theory, the diffraction model, derives from the simple model proposed by Ziman [5] for liquid metals. In this model the electrons are assumed to propagate with well defined k vectors and the scattering by the ions is handled by the golden rule, multiple scattering being neglected. The resistivity is then calculated from the Boltzmann equation. For a single component liquid or amorphous metal one finds

$$\rho \propto \int_0^{2k_F} a(q)\,|t(q)|^2\, q^3 dq \qquad (3)$$

where k_F is the Fermi wave vector.
The structure factor, a(q), results from a sum over the position of the ions with factor $e^{i\vec{q}.\vec{r}}$; $|t(q)|^2$ is the square of the t-matrix for scattering through a wave vector q; of the factor q^3, q^2 comes from the volume element of k-space and q from the scattering angle factor (1-cosθ).

This very simple expression has been used to account for the cor-

relation of the TCR with ρ in the following way:
In (3), the factor q^3 implies that ρ will be dominated by the integral at $2k_F$. If $a(q)$ is the most rapidly varying function with q, then ρ will vary roughly as a $(2k_F)$. As the temperature is raised a(q) smooths out; thus systems with large $a(2k_F)$, and hence large ρ, will have a negative TCR while those with small $a(2k_F)$, and hence small ρ, will have a positive TCR. In fact the situation is not quite so simple because we have neglected contributions from the dynamic part of the structure factor-i.e. inelastic scattering due to thermal vibrations which gives a positive contribution to ρ. Furthermore all amorphous metals are in fact alloys so that the sum over the ions results in partial structure factors with different matrices for each atom type[6]. As a result ρ and the TCR are a little more complicated than suggested by equation (3) but it is still true, a large ρ and negative TCR result when $2k_F$ is near the peak of some average structure factor (i.e. the structure factor suitably weighted by the different t-matrices [7]).

Now let us turn to the thermopower. The Mott expression, equation (2), (valid in metallic glasses because phonon drag is absent) is often written for simple liquid metals in the form

$$S(T) = -\frac{\pi^2 k_B^2 T}{3|e|E_F}\left(3-2q-\frac{1}{2}r\right) \tag{4}$$

where

$$q \propto |t(q=2k_F)|^2 \, a(2k_F) \tag{5}$$

$$r \propto \int_0^{2k_F} a(q)\left|k_F \frac{\partial}{\partial q}|t(q)|^2\right| q^3 dq \tag{6}$$

The r term is often ignored.

Equation (5) has been argued to give a correlation [8] between S and ρ similar to that between $\frac{d\rho}{dt}$ and ρ. The resistivity is large when q is large and if q is large enough S will be positive; whereas when ρ is small, q is small and so S will be negative.

How does the diffraction model work? For simple glasses the evidence is that it works well, not just qualitatively but quantitatively. For example in $Ca_{70}Mg_{30}$, Hafner and Phillip [9] have calculated both the resistivity and its temperature dependence, using appropriate pseudopotentials, partial structure factors and Debye temperature. They calculate a resistivity of 35.6 μΩ-cm compared with a measured value of 43.7 μΩ-cm and find good agreement with the measured temperature dependence (though it must be admitted that the data are somewhat scattered). The agreement, shown in figure 4, is made more impressive by the fact that there are no adjustable parameters. Similar good agreement was found by Coté and Meisel [10] for the resistivity of $Mg_{70}Zn_{30}$. Since the TCR of $Ca_{70}Mg_{30}$ and $Mg_{70}Zn_{30}$ are of opposite sign, these two calculations effectively include the Mooij correlation [11].

Thermopower is more sensitive than resistivity to the values of the parameters and, perhaps because of this, detailed agreement between the diffraction model and the measured thermopower has yet to be given. Figure 5 shows the composition and temperature dependence of the thermopower in Mg-Zn. The magnitude of S can be approximated from eqn. (4) using ρ to estimate q; the results are good. The composition dependence is quite consistent with the diffraction model but an explanation of the temperature dependence (which is obviously non/linear) requires a more detailed knowledge of partial structure factors than is generally available. Clearly there is more work to be done here, but there is no reason to doubt the validity of the diffraction model for these simple alloys.

But most amorphous metals are not simple; they have an open d-band. How does the diffraction model work here? Qualitatively it can describe some properties quite well - for example there is obviously a Mooij correlation! Also one can sometimes find good correlations between S,ρ and the TCR, as outlined above. An example is shown in figure 6. However when one tries to make a quantitative calculation, difficulties arise. This is well illustrated by calculations for the resistivity of liquid (or amorphous) iron. Calculations lead to values anywhere between 276 μΩ-cm and 1100 μΩ-cm [12] whereas the measured value is only 135 μΩ-cm. These discrepancies result in part from the extreme sensitivity of the calculations to the exact location of E_F. But there are also more fundamental objections to using the diffraction model for d-band amorphous alloys. These are:

a) with a short mean free path , the k-vector of the conduction electron is no longer defined.

b) multiple scattering effects become important.

3) at high resistivities there are contributions to the conductivity from the d-band itself.

To get around these problems an alternate method has been tried, principally by Ballentine [13,14]. The method is called LCAO recursion method. It starts with a numerical LCAO calculation of the density of electron states for a random atomic cluster generated by computer. The spectral density of electron states is evaluated by a recursion technique. For the method to work the cluster must be larger than the mean free path, which effectively limits the method to systems with a very short mean free path. Under such circumstances the conductivity is really a diffusion process so that equation (1) becomes

$$\sigma = e^2 N(E_F) D(E_F) \qquad (7)$$

$D(E_F)$ being the diffusion constant at E_F. Ballentine evaluates $D(E_F)$ numerically using the Kubo method with a similar recursion technique[14]. There is no requirement for the electrons to be labelled by a k-vector and no a priori assumptions are made about whether s- or d- electrons carry the current. The total number of valence electrons (s and d) is of course known, and this number then determines E_F, $N(E_F)$ and $D(E_F)$. Unfortunately the fact that the method is numerical prevents one from seeing qualitative features as one can for the diffraction method.

However the results are rather striking as is shown in figure 7. Calculations of the resistivity of liquid (or amorphous) transition metals are in excellent agreement with the measurements. The small differences could just be due to thermal vibrations or magnetic effects. One of the surprising results of this calculation is that the d-electrons dominate the conduction process in many of these materials - contributing over 90% in Fe & Mn. This is in strong contrast to the assumptions of the diffraction model which assumes that conduction is by s-electrons alone. By itself this result can explain why the diffraction model breaks down. It also leads one to anticipate a different dependence of σ on the density of states. If d-electrons dominate the conduction process then σ wll <u>increase</u> with $N(E_F)$ (though there may be opposing effects from $D(E_F)$); in the diffraction model σ <u>decreases</u> as the density of d-states increases because the s-electron lifetime is controlled by s-d scattering.

The LCAO recursion method is still in its infancy so far as metallic glasses are concerned: no calculations exist yet for thermopower, TCR or concentration dependence. But these initial successes give one to hope that it may well be the answer to the difficulties of handling metallic glasses with open d-bands. The method is complementary to the diffraction model. When the d-density of states gets too low the current becomes dominated by s-electrons whose mean free path is now too long for the LCAO recursion method to be practical. But now the diffraction model would become applicible. At a rough guess the change over between the two models appears to be about 80 μΩ-cm; and it may well be that the concentration dependence of the resistivity of Ni-Zr shown in figure 6 is no more than this transition.

3. THE HALL EFFECT

According to the diffraction model, the Hall effect in amorphous metals is extremely simple: the Hall coefficient R_H, is negative, independent of temperature and equal to the free electron value. Table 2 shows the measured Hall effect for some simple metallic glasses together with calculated values from the free electron model; clearly there is excellent agreement which is further strong evidence for the validity of this model for simple systems. The open d-band metallic glasses on the other hand are much more complicated. The Hall coefficient is <u>positive</u> in many systems and there can be a strong composition dependence somewhat reminiscent of the thermopower and T.C.R. Data for Zr-glasses as shown in figure 8.

For these d-band alloys no satisfactory explanation has been given of the positive Hall effect, and one can say little more than it is further proof that the diffraction model is not applicable to these alloys. Recent work by Harris [15)] suggests that a positive Hall effect can result from a strong energy dependence to the scattering which in fact implies a correlation of positive Hall effect with positive thermopower. The correlation is not fully realized. An alternate approach by Cochrane et al.[16] has related the positive Hall coefficient to the side-jump effect[17] normally seen only in ferromagnets but possibly visible in strongly paramagnetic amorphous alloys like Zr-Fe because of the high resistivity. Again this conclusion must remain tentative. In all likelihood the situation will remain unresolved until the LCAO recursion method is applied to the Hall effect.

4. LOW TEMPERATURE RESISTIVITY & THERMOPOWER

Let us now turn to some special effects at low temperatures, below about 10K, starting with the resistivity. The basic fact is simply stated: at low temperatures $\frac{\partial\rho}{\partial T}$ is always negative[18], irrespective of its value at high temperature except, of course, for superconducting alloys at or close to T_c. This is illustrated in figures 9 and 10 for three alloys. This seemingly universal result is quite different from the Mooij correlation and leads one to believe it has a different physical origin. A related and equally interesting result is that the magnetoresistance, which is essentially zero at high temperatures in nonmagnetic amorphous alloys, becomes anomalously large at low temperatures and has a characteristic shape[19], shown in figure 11, varying roughly as H^2 at low field and as $\sqrt{H}$ at high field. None of these results are explicable within the framework of the diffraction model.

At the lowest temperatures, say below 4K, the contribution to the temperature dependence of ρ from changes in the structure may be neglected and one is left with what may be considered an intrinsic variation. Many different characterizations of this dependence have been offered in the past (see e.g. Harris & Strom-Olsen)[20] but the simplest characterization is that ρ varies as $\sqrt{T}$.[18] This is shown in figure 12. The variation with $\sqrt{T}$ in fact provides the clue to the possible origin of the behavior, since it is predicted by various general theories of electron conduction in highly disordered conductors[21].

In the limit of very intense scattering, multiple scattering effects lead to two prominent quantum corrections to the conductivity.

These corrections are generally known as weak localization and coulombic ineraction. Weak localization is a one-electron interference effect (Bergmann)[22] resulting from electrons being scattered from k to -k by two complementary sequences (viz : $\vec{k} \to \vec{k}-\vec{k}_1 \to \vec{k}-\vec{k}_1-\vec{k}_2 \ldots\ldots \to \vec{k}-\vec{k}_1-\vec{k}_2-\ldots-\vec{k}_n = -\vec{k}$ and $\vec{k} \to \vec{k}-\vec{k}_n \to \vec{k}-\vec{k}_n-\vec{k}_{n-1} \to \ldots\ldots \to \vec{k}-\vec{k}_n-\vec{k}_{n-1} \ldots\ldots -\vec{k}_1 = -\vec{k}$).If <u>all</u> the scattering events are elastic then two sequences have exactly the same phase change and the two back-scattered waves interfere constructively giving twice the scattering intensity. However if an inelastic scattering process intervenes the phase coherence is destroyed and the effect is removed. Thus the effect is only seen when the elastic scattering is very intense - i.e. in disordered systems at temperatures low enough for inelastic electron-phonon and electron-electron scattering process to be insignificant. In amorphous metals this really means low temperatures as we have defined it. The second quantum correction effect, coulombic interaction, is a many-body effect and is principally deterioration of screening due to intense scattering - the screening hole cannot follow the electron effectively. This results of course in an increase in resistivity. This effect too decreases with inelastic scattering. Thus both W-L and coulombic lead <u>in the simplest case</u> to an increase of conductivity or a decrease of resistivity as the temperature increases. (More complicated effects occur when, for example, spin-orbit or spin-scattering is present).

Approximately one finds due to W-L

$$\rho(T) - \rho(0) \sim -\tau_{in}^{1/2} \sim -T \quad \text{for inelastic electron-phonon scattering} \qquad (8)$$

and due to C-I $\quad \rho(T) - \rho(0) \sim -\sqrt{\frac{T}{D}} \qquad (9)$

The fact that $\rho \sim \sqrt{T}$ suggests that the coulombic interaction effect is dominant in determining the low temperature dependence of ρ and indeed when the actual parameters are put in these models this suspicion is confirmed. For example in Y-Al[19] one finds that, for reasonable parameters, the W-L contribution is negligible below 4K and that the C-I contribution $\sqrt{T}$ slope gives $D \sim 8\times10^{-5}\ m^2 s^{-1}$, as compared with the figure of $D \sim 10 \times 10^{-5}\ m^2 s^{-1}$ from the absolute value of the conductivity. Similar results obtain for other amorphous metals [23]. The situation with the magneto-resistance is not quite so satisfactory but still encouraging. A magnetic field causes dephasing effects similar to those of inelastic scattering, which implies that the size of the magneto-resistance at low temperatures is similar to that of, say the change in $\rho(T)$ between 10K and 1K - which is in line with experimental results. It also says that the magnetoresistance should disappear when dephasing effects are already accomplished by inelastic scattering - i.e. at higher temperatures - as is also observed. Both W-L and C-I predict a magnetoresistance $\sim H^2$ at low field going to $\sqrt{H}$ at high field, also as observed, but it turns out this time that, at the lowest temperatures, the contributions from W-L generally dominate over C-I. When one comes to make a quantitative fit, there are some discrepancies - perhaps a factor or two between theory and experiment [19].

It should be noted that in their current state the theories have the same difficulties as the diffraction model: namely that they require the electrons to propagate with well defined k-values and so cannot be expected to give exact quantitative agreement with experiment. It is to be hoped that this deficiency will be remedied over the next few years.

By contrast with the resistivity the low temperature thermopower shows fewer surprises. The thermopower in many non magnetic glasses is roughly linear with temperature from room temperature down. The one feature of real interest is a change of slope which typically occurs round about 50K. This "knee", which is clearly visible in figure 1, has been interpreted as due to a change in the electron effective mass due to phonons [24]. At low temperature the electrons carry around a cloud of phonons which increases the effective mass from m_o to $m_o(1+\lambda)$ where m_o is the band mass and λ is the electron phonon coupling constant which appears in superconductivity. At high temperatures thermal vibrations destroy this cloud and the effective mass returns to the band mass m. The resulting effect on the thermopower [25] is that the ratio

$$\frac{(S/T)_{lowT}}{(S/T)_{highT}} = 1 + \lambda(o) \qquad (10)$$

This effect is in principle present in all metals, but is generally masked by the phonon drag contribution to the thermopower. In amorphous metals the electron and phonon heat currents relax directly via the disorder so phonon-drag vanishes and the effect of eqn. (10) becomes visible. Analysis on systems like Ni-Zr[26] which become superconducting (allowing an independent determination of λ) support this interpretation. If correct it could prove a powerful tool for determining the electron phonon coupling constant in non superconducting glasses.

5. MAGNETIC AMORPHOUS METALS : EXAMPLES OF SPECIAL EFFECTS

Historically the first amorphous alloys whose transport properties were measured in any quantity were magnetic, which was unfortunate because it was difficult to be sure which effects resulted from the atomic disorder and which from the magnetism. In some cases a property may be almost unaffected by magnetic order, in others it may be dominated by it. This is illustrated in figure 13 by the resistivity of $(Fe_{1-x}Cr_x)_{.8}B_{.2}$ alloys. $Fe_{80}B_{20}$ shows a behavior characteristic of non-magnetic d-band amorphous alloys (e.g. $Ni_{75}P_{25}$) but the addition of just a small amount of Cr produces a second minimum which is in fact of magnetic origin [27] and which disappears only when sufficient Cr is added to destroy the magnetic moment. In this example it happens that the magnetic anomaly reflects the amorphous state and would not be seen in the crystalline state. In the case of Zr-Fe [28], figure 14, the bump in the resistivity is seen on either side of the critical composition for ferromagnetism and it caused by spin fluctuations. A similar anomalies are seen in crystalline alloys but the effects are made more dramatic by being superposed on a negatively sloped background instead of a positively sloped one as would be the case for crystalline systems.

Thermopower may also show some markedly different behavior in magnetic amorphous metals compared with the relatively simple behavior in non-magnetic systems. Figure 15 shows the behavior in Fe-B where the large bump would appear to be related to the occurence of magnetism - though no quantitative account has yet been given.

6. CONCLUSIONS

Non magnetic amorphous alloys are conveniently divided into two classes : simple alloys with no d-band (or with a very low d-density of states) and d-band alloys. Simple alloys generally have a resistivity below 100 μΩ-cm and, so far can be determined, all their transport properties above 10K are well described by the diffraction model. d-band alloys (which numerically constitute by far the greater fraction) generally have resistivities above 100 μΩ-cm and have many properties which are in direct contradiction to the diffraction model. For these alloys a promising new approach is the LCAO recursion method, though much work remains to be done before the full range of their properties is understood. At low temperature all glasses show properties which are controlled by quantum correction effects due to the intense scattering. These effects - weak localization and coulombic interactions - control the resistance and magnetoresistance below about 10K.

REFERENCES

1) Ashcroft, N.W. and Mermin, N.D.in Solid State Physics (Holt, Rinehart Wilson, New York, 197).
2) Mott, N.F. and Jones, H. in The Theory and Properties of Metals and Alloys (Dover Press, New York, 1958).
3) Mooij, J.H., Phys. Status Solidi, A 17, 521 (1973).
4) Mitzutani, U., Proceedings of Fifth International Conference on Amorphous Metals, RQ-5, Würzburg, ed. Steeb and Warlimont, H. (North-Holland 1985), 975.
5) Ziman, J.M., Philos Mag.6, 1013 (1961).
6) Faber, T.E. and Ziman, J.M., Philos. Mag. 11, 153 (1965).
7) It should be noted that it is dangerous to take the peak of a(q) for these purposes from a single X-ray diffraction pattern since the structure factor is averaged quite differently. Though such comparisons are common in the literature they are generally meaningless.

8) Nagel, S.R., Phys. Rev. Lett. 41, 990 (1978).

9) Hafner, J. and Phillip, F., J. Phys. F: Metal Physics 14, 1685 (1984).

10) Cote, P.J. and Meisel, L.V., Proceedings of the Fifth International Conference on Liquid & Amorphous Metals, LAM V, Journal of Non-Crystalline Solids 61 and 62, 1167 (1984).

11) Hafner and Phillip point out that although most of the Mooij correlation in simple glasses can be explained by the position of $2k_F$ with respect to $a(\vec{q})$, detailed comparison requires inclusion of inelastic scattering effects as well.

12) Evans, R., Greenwood, D, Lloyd, P., Phys. Lett. A 35, 57 (1971); Esposito, E. Ehrenrich, H., Phys. Rev. B 18, 3913 (1978).

13) Ballentine, L.E., Proceedings of the Fifth International Conference on Rapidly Quenched Metals (RQ-5) Wurzburg, 1985, ed. Steeb, S. and Warlimont, H. (North Holland), 977.

14) Bose, S.K., Ballentine, L.E. and Hammerberg, J.E., J. Phys. F. 13. Metal Physics 2089 (1983). Ballentine, L.E. and Hammerberg, J.E., Can. J. Phys. 62, 692 (1984).

15) Harris, R., J. Phys. F: Metal Physics (1985, in press).

16) Cochrane, R.W., Trudeau, M. and Strom-Olsen, J.O., J. Appl. Phys. 55, 1939 (1984).

17) Berger, L., Phys. Rev. B2, 4559 (1970).

18) Cochrane, R.W. and Strom-Olsen, J.O., Phys. Rev. B29, 1088 (1984). (The $\sqrt{T}$ dependence in metallic glasses was first reported by Rapp, O., Baghat, S.M. and Gudmundsson, H., Solid State Commun. 42, 741 (1982).

19) Olivier, M., Strom-Olsen, J.O., Altounian, Z. and Cochrane, R.W., Proc. LITPIM Supp., 55 (PTB Braunschweig 1984).

20) Harris, R. Strom-Olsen, J.O., in Glassy Metals II (Springer-Verlag 1982) 325.

21) A good recent review is: Lee, P.A., and Ramakrishnan, T.V., Rev. Mod. Physics 57, 287 (1985).

22) Bergman, G., Phys. Reports.

23) Bieri, J.B., Fert, A., Creuzet, G. and Ousset, J.C., Solid State

Communications 49, 849 (1984); Poon, S.J., Wong, K.M. and Drehman, A.J., Phys. Rev. B 31, 1668 (1985).

24) Gallagher, B.J., J. Phys. F 11, L207 (1981).

25) Kaiser, A.B., Phys. Rev. B 29, 7088 (1984).

26) Altounian, Z. and Strom-Olsen, J.O., Phys. Rev. B 27, 4149 (1983).

27) Olivier, M. and Strom-Olsen, J.O. (unpublished).

28) Strom-Olsen, J.O., Olivier, M., Altounian, Z., Muir, W.B. and Cochrane, R.W., Proceedings of the Fifth International Conference on Liquid and Amorphous Metals (LAM V), J. Non Crystalline Solids 61 and 62, 1391 (1984).

29) Meaden, G.T., Electrical Resistance of Metals (Heywood, London, 1965).

30) Huebner, R.P., Phys. Rev. 171, 634 (1968).

31) Altounian, Z., Foiles, C-L., Muir, W.B. and Strom-Olsen, J.O., Phys. Rev. B 27, 1955 (1985).

32) Baibich, M., Muir, W.B., Altounian, Z. and Tu Guo-Hua, Phys. Rev. B 26, 2693 (1982).

33) Matsuda, T. and Mizutani, U., Solid State Commun. 44, 145 (1982).

34) Mizutani, U. and Matsuda, T., J. Phys. F: Metal Physics 13, 2115 (1983).

35) Richter, R., Altounian, Z. and Strom-Olsen, J.O., J. Mat. Sci. Lett. 4, 1005 (1985).

36) Buschow, K.H.J., J. Phys. F: Metal Physics 13, 563 (1983).

37) Altounian, Z., Tu Guo-hua and Strom Olsen, J.O., J. Appl. Phys. 54, 3111 (1985).

38) Strom-Olsen, J.O., Olivier, M., Altounian, Z., Muir, W.B. and Cochrane, R.W., J. Non Crys. Solids 61 and 62, 1391 (1984).

Table 1:

Resistivity and thermopower of selected amorphous metals

	ρ ($\mu\Omega$-cm)	$\frac{1}{\rho}\frac{d\rho}{dT}$ ($10^{-4}K^{-1}$)	S ($\mu V\ K^{-1}$)	Mass Density (g cm^{-3})	Electron Density ($10^{23}cm^{-3}$)	Open d-band?	References
$Ca_{65}Al_{35}$	400	-4.4	+2.0	1.90	0.76	Yes	8,34
$Y_{70}Al_{30}$	260	-1.0	----	4.10	1.05	Yes	35
$Ti_{70}Ni_{30}$	215	-2.5	----	5.54	3.78	Yes	36
$Ni_{60}Zr_{40}$	175	-0.66	+2.0	7.81	4.98	Yes	31, 37
$Fe_{80}B_{20}$	120	+1.27	-3.4	7.51	6.17	Yes	38
$Mg_{80}Cu_{20}$	56	-1.7	----	----	----	No	33
$Mg_{70}Zn_{30}$	50	-0.18	~ 0	2.92	0.96	No	32
$Ca_{70}Mg_{30}$	44	+1.1	----	1.6	0.54	No	9

Table 2:

Hall coefficient in three simple amorphous alloys.
Data from Mizutani and Yoshida, J. Phys. F 12, (1982).

	R_H measured	R_H calculated
$Ag_{.5}Cu_{.5}$	-9.11 ± 0.13	-9.16
$(Ag_{.5}Cu_{.5})_{.5}Mg_{.5}$	-6.60 ± 0.13	-6.73
$(Ag_{.5}Cu_{.5})_{.5}Al_{.5}$	-4.62 ± 0.26	-4.50

Units: $10^{-11}\ M^3A^{-1}\ S^{-1}$

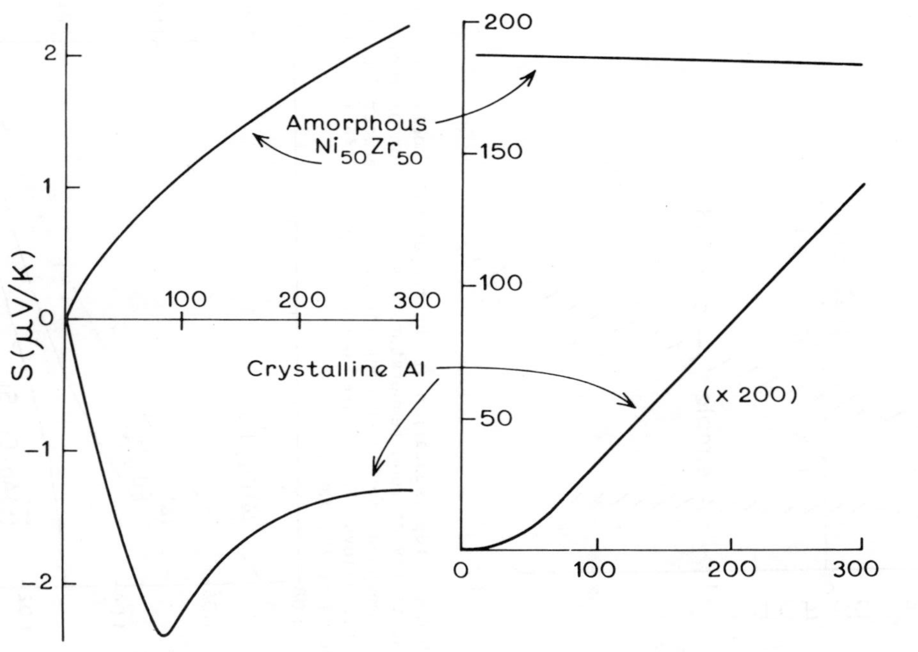

Fig.1. The resistivity of crystalline Al (Ref.29) together with its thermopower (Ref.30), contrasted with the same properties in amorphous $Ni_{50}Zr_{50}$ (Ref. 31).

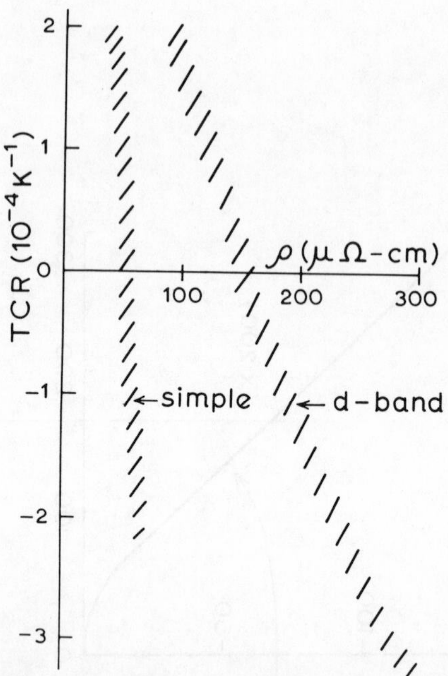

Fig.2 Schematic representation of the temperature coefficient of resistivity at room temperature as a function of resistivity for amorphous metals showing the division between d-band and simple alloys. The shading gives an approximate idea of the spread of the data.

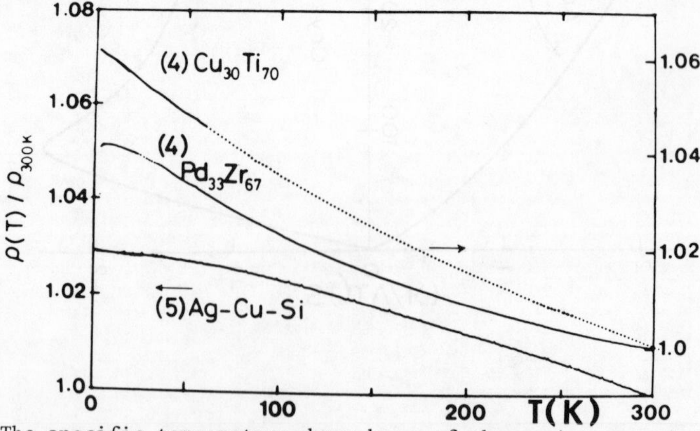

Fig.3. The specific temperature dependence of the resistance of one simple metallic glass (Ag-Cu-Si) and two d-band glasses ($Cu_{30}Ti_{70}$ and $Pd_{33}Zr_{67}$). Note the opposite curvature. Data from reference 4.

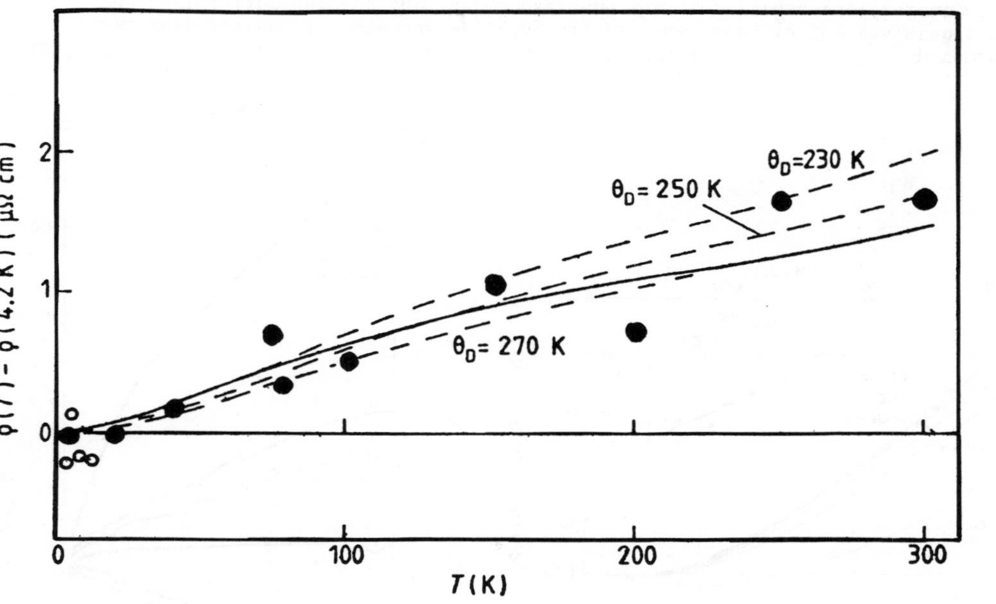

Fig.4. The calculated and measured resistivity of $Ca_{70}Mg_{30}$ (Ref. 9). The calculations are detailed applications of the Faber-Ziman expression (Ref.6) with all parameters taken from independent sources. The solid curve using the measured vibration spectrum by neutron diffraction; the dashed curves are the Debye approximation with θ_D as indicated. Experimental values are solid points.

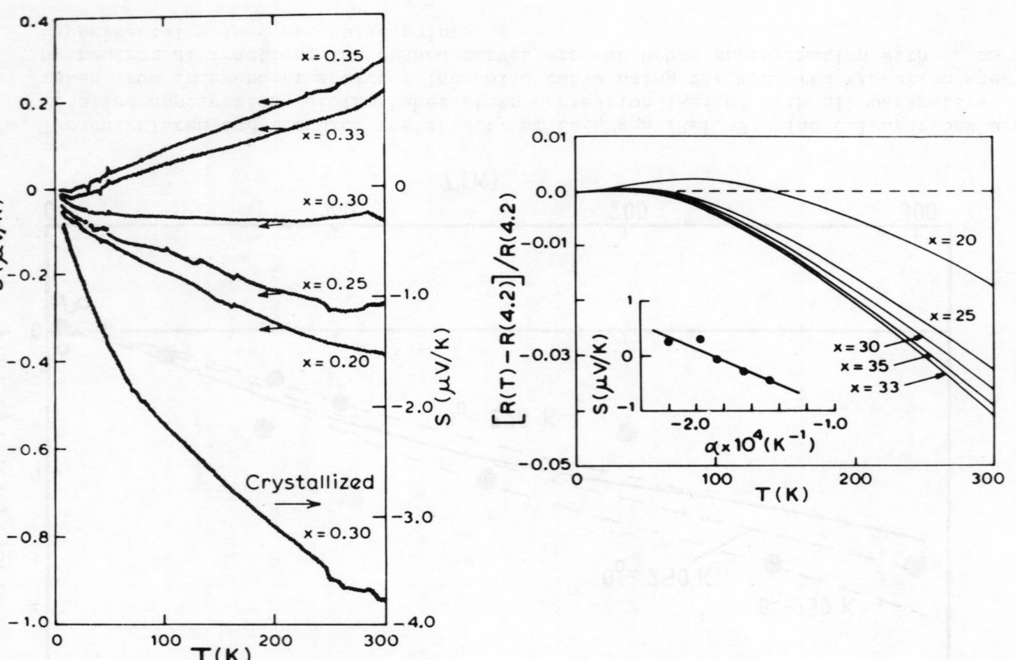

Fig.5. The dependence of thermopower, S, and electrical resistance, R(T), on temperature and composition in amorphous $Mg_{1-x}Zn_x$ alloys. The inset to the resistance plot shows the correlation between the thermopower and the temperature coefficient of resistivity. Data from reference 32.

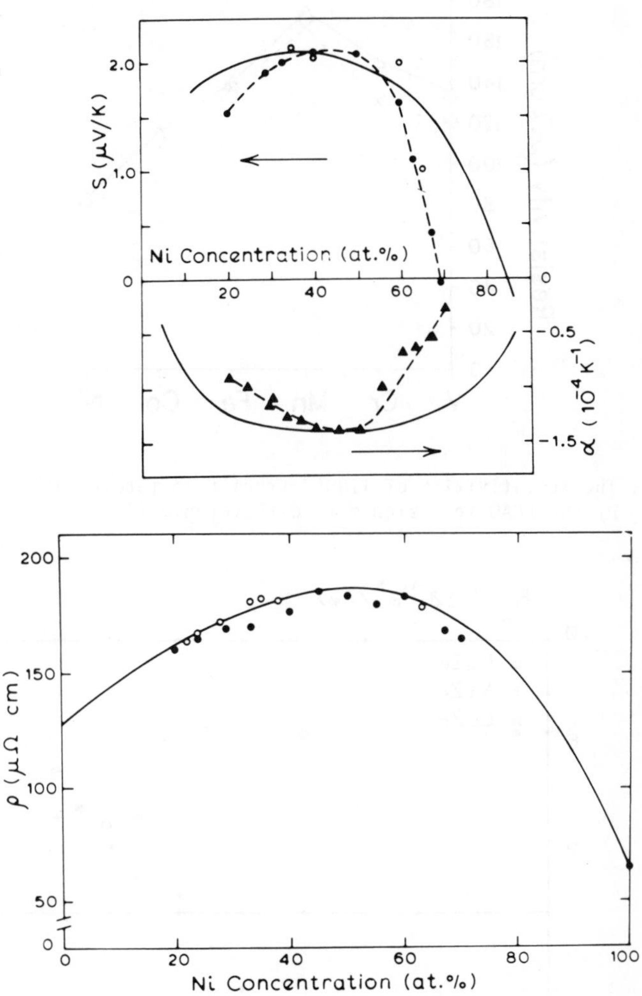

Fig.6. The thermopower, S, temperature coefficient of resistivity, α, and resistivity ρ in amorphous Zr-Ni. Note the correlation in composition dependence of all three quantities. The solid lines represent a fit of the composition dependence to the diffraction model. From reference 31.

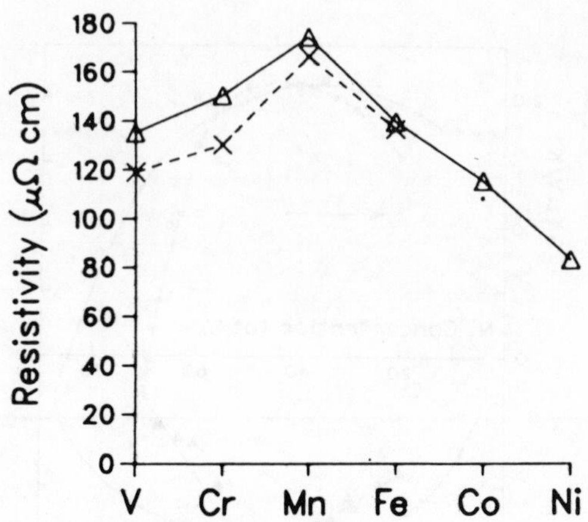

Fig.7. The resistivities of liquid transition metals calculated by the LCAO recursion method (Reference 13)

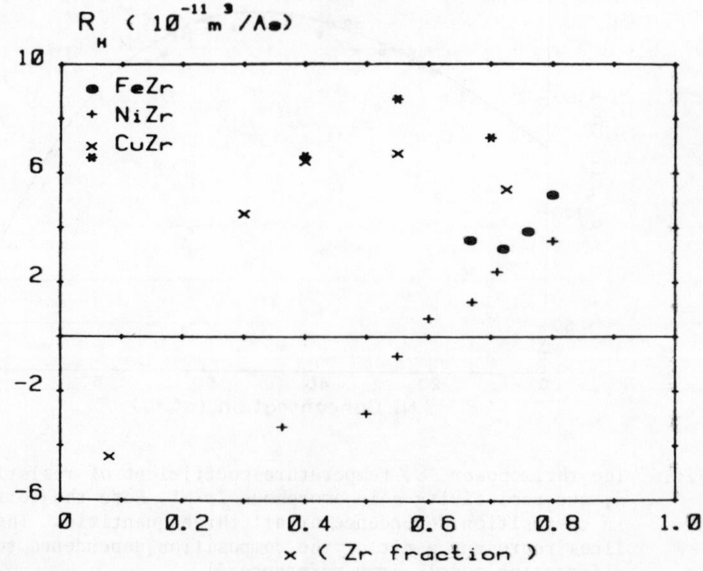

Fig.8. Hall coefficients in amorphous Zr-based alloys (Reference 16).

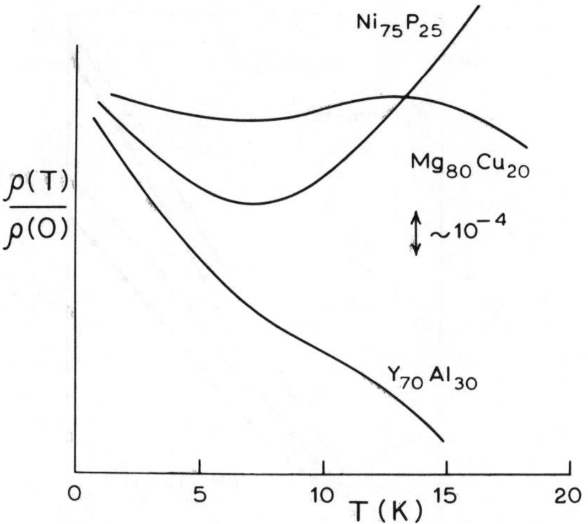

Fig.9. Schematic of the temperature dependence below 20K of the electrical resistivity of three amorphous alloys.

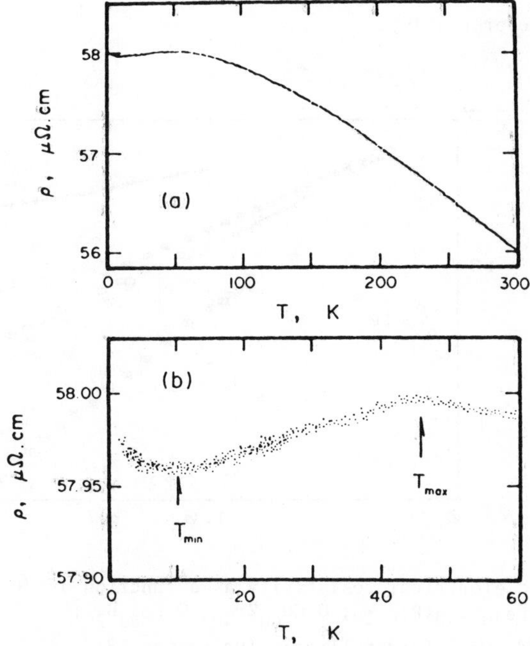

Fig.10. Resistivity of $Mg_{80}Cu_{20}$. Data from Reference 33.

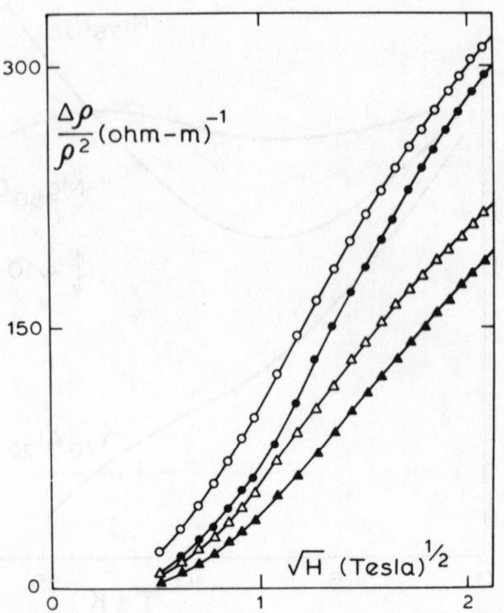

Fig.11. Magnetoresistance in amorphous Y-Al. Circles $Y_{80}Al_{20}$, triangles $Y_{60}Al_{40}$. Open symbols 1.2K, filled symbols 4.2K. (Reference 19).

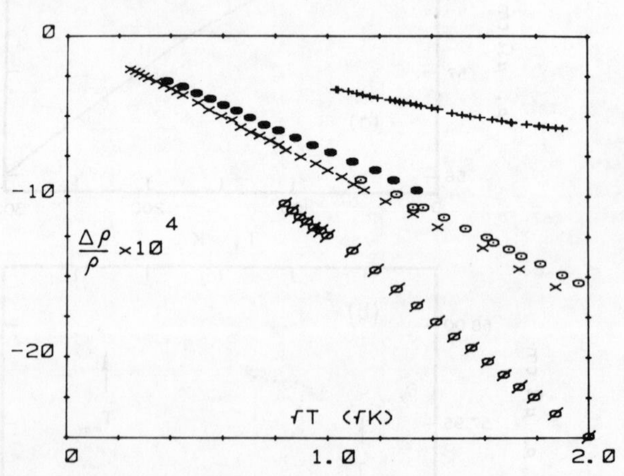

Fig.12. The electrical resistivity as a function of $\sqrt{T}$: ∅ $Y_{77.5}Al_{22.5}$; X $Fe_{40}Ni_{40}(P,B)_{20}$; O $Cu_{60}Zr_{40}$; O $Fe_{80}B_{20}$; +$Pd_{82}Cr_{18}$ (crystalline). (Reference 18).

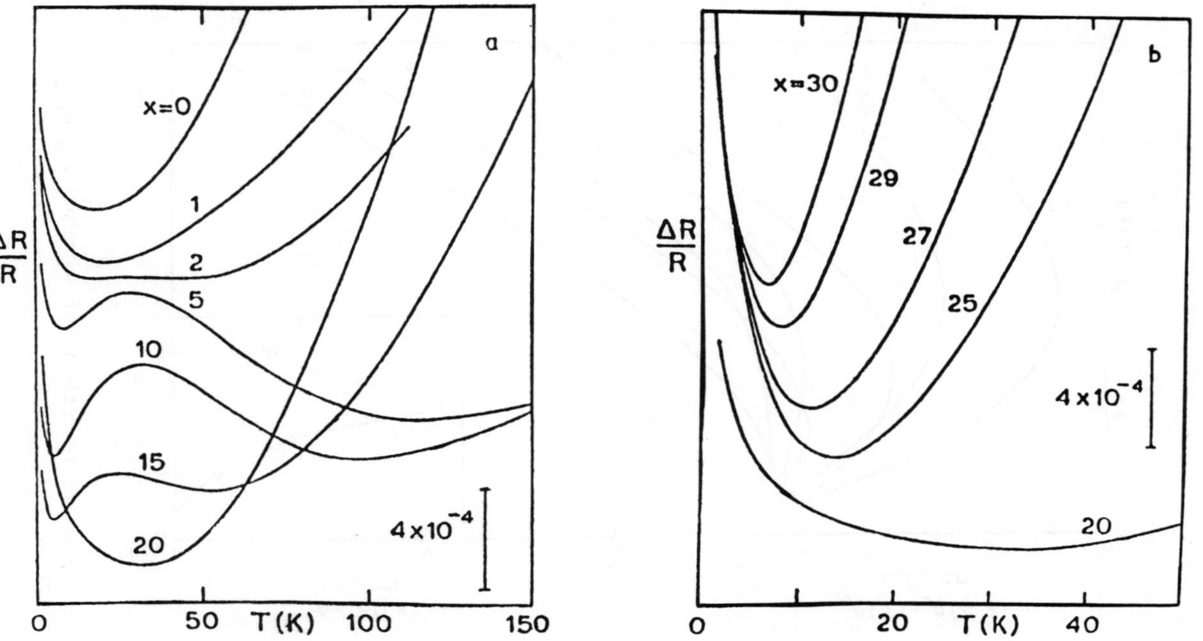

Fig.13. The electrical resistivity of amorphous $(Fe_{100-x}Cr_x)_{.8}B_{.2}$ (Reference 27).

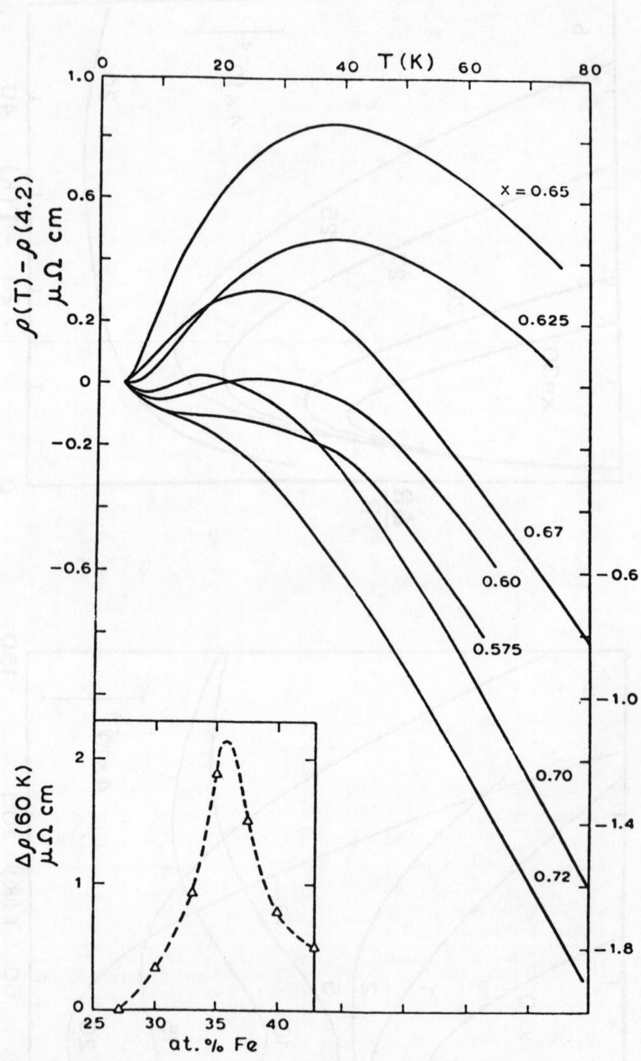

Fig.14. The electrical resistivity of amorphous Zr_xFe_{1-x} (Reference 28).

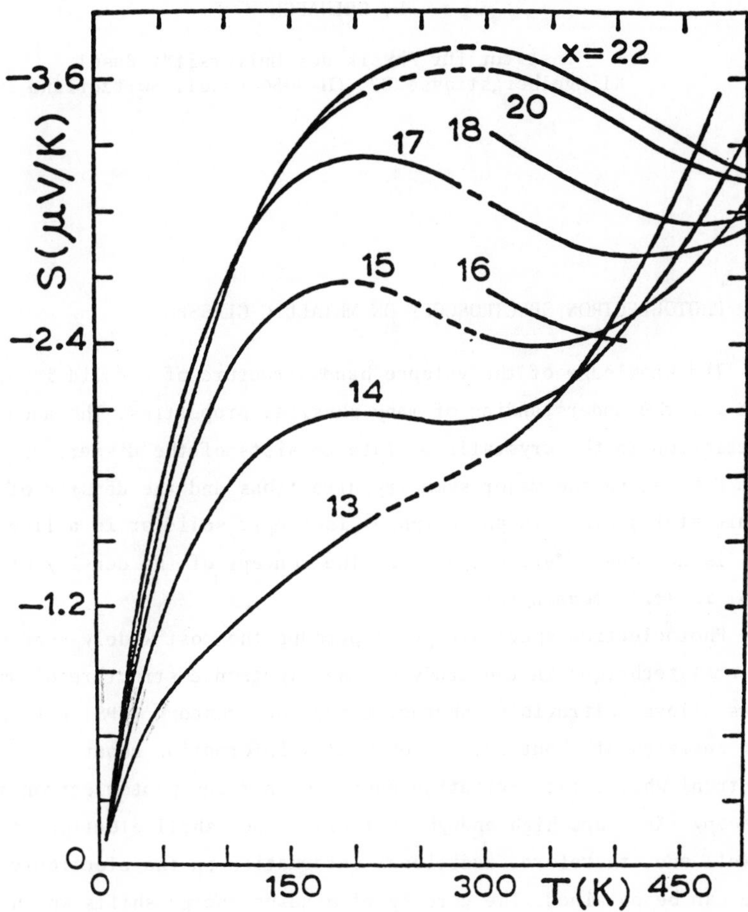

Fig.15. The thermopower of amorphous $Fe_{100-x}B_x$.

ELECTRONIC STRUCTURE OF METALLIC GLASSES

P. OELHAFEN

Institut für Physik der Universität Basel
Klingelbergstrasse 82, CH-4056 Basel, Switzerland

1. PHOTOELECTRON SPECTROSCOPY ON METALLIC GLASSES

The knowledge of the valence band structure of a solid is significant for the understanding of many physical properties. The adequate description in the crystalline state consists of the dispersion relation $E(\underline{k})$ along the major symmetry directions and the density of electronic states $D(E)$. In an amorphous isotropic solid or in a liquid, $E(\underline{k})$ is no longer defined, whereas the concept of the density of states (DOS) is still meaningful.

Photoelectron spectroscopy is perhaps the most widely used experimental technique in the study of the electronic structure of amorphous alloys. Ultraviolet photoelectron spectroscopy (UPS) with excitation energies of about 20...40 eV yields information about the valence electrons whilst the excitation energies in X-ray photoelectron spectroscopy (XPS) are high enough to ionize inner shell electron states as well and, therefore, additional information on the electronic structure can be obtained. The core level binding energy shifts which are related to the charge rearrangement on alloying and the changes of the core line asymmetries, which can yield at least qualitative information about the local density of states, are of particular interest here. XPS core level spectroscopy also provides important information in the form of a chemical analysis of the sample surface within the probing depth of photoelectron spectroscopy which varies between about 0.5 and 2 nm depending on the kinetic energy of the excited electrons[1).

An important question in connection with UPS valence band spectroscopy on amorphous alloys concerns the problem of how far the spectra reflect the density of initial states i.e. the DOS below the Fermi level E_F. In the crystalline case it is well known that UPS valence band spectra can deviate appreciably from the density of states for several reasons: (i) the final state of the excited electron is influenced by the DOS above E_F and (ii) due to $\underline{k}$-conservation the direct optical transitions lead to structures in the energy distribution of the emitted photoelectrons which are not related to DOS effects. The first point becomes irrelevant as soon as the excitation energy is high enough and the final state is located in a free electron like continuum well above E_F. The second point only holds for the crystalline case since $\underline{k}$-conservation is no longer an important selection rule in the amorphous or liquid state.

In practice the influence of the empty electron states can be checked by measuring valence band spectra with different excitation energies, leading to different final states. This point is illustrated in Fig.1, showing valence band spectra of amorphous $Pd_{27}Hf_{73}$ measured with different photon energies: 21.2, 40.8 and 1486.6 eV [2)]. It is obvious from Fig.1 that the general shape of the three spectra is very similar with respect to peak widths and positions. Therefore, the spectra are essentially determined by the initial (occupied) density of states and the final (empty) state density, which has been varied in Fig.1 by using different excitation energies, does not influence the shape of the spectra. The main difference in the spectra is the relative intensity of the two pronounced peaks. As will be discussed later on, the peak near the Fermi edge contains mainly Hf 5d-electron states whereas the Pd 4d-electron states are essentially located in the peak at a binding energy of about 4 eV. Since the photoelectron excitation cross sections for the Hf 5d and the Pd 4d-electron states depend in a different way on the excitation energy the relative peak intensity is changing as a function of energy as well. In fact, calculated atomic subshell photoionization cross sections [3)] yield a cross section ratio for Pd 4d to Hf 5d electron states of 2.4 and 12.6 for HeI and HeII excitation, respectively. This tendency is clearly visible in Fig.1 since the intensity of the "Pd peak" (near a binding energy of 4 eV) is increasing with respect to the "Hf peak" (close to E_F) by increasing the photon energy from 21.1 to 40.8 eV. The peak at 6 eV in the valence band spectrum measured with 40.8 eV is due to a minor oxygen contamination of the sample surface and is not visible in the other two spectra since the surface sensitivity is most pronounced in the case of an excitation energy of 40.8 eV.

Moreover the similarity of the three spectra shown in Fig.1 indicates that the data reflect more bulk than surface properties of the alloys. The change in the escape depth of the photoelectrons of spectra measured with 40.8 and 1486.6 eV is of the order of a factor of two. Since the biggest change in the electronic structure is expected to occur within the first two atomic layers at the surface, we can conclude that even the spectrum measured with 40.8 eV, where we are close to the minimum in the escape depth, reflects essentially a bulk DOS if we leave out of consideration the peak at about 6 eV which is due to oxygen contamination. An additional argument for the relevance of UPS valence band spectra to obtain bulk information will be given by comparing photoelectron spectra from clean glassy alloy surfaces with band structure calculations.

The stability model by Nagel and Tauc [4)] is an interesting aspect in the context of the valence band density of states of amorphous alloys. Within this model the valence electrons are considered to be nearly free and an increased stability against crystallization has been found when the Fermi level E_F is located at a minimum in the DOS. The size and location of this minimum is in turn determined by the structure factor of the system. For an amorphous system with a spherically symmetric structure factor the system will be in a metastable state. Therefore, many photoemission experiments have been performed on amorphous alloys in order to investigate the DOS behaviour near E_F [5)]. However, this kind of study is restricted to nearly free electron-like systems since the electronic structure in transition metal alloys is dominated by d-electrons and therefore a nearly free electron model is no longer applicable. A minimum in the DOS at E_F, which is obviously related to the model of Nagel and Tauc, has been observed in different amorphous alloys such as quench condensed films of noble metal-tin alloys [6], glassy Ca-Al [7] and Mg-Zn [8]. However, we are not going to discuss these systems in more detail. We will mainly focus on glassy transition metal alloys for which we will review the general common valence band properties in order to understand their relation to the chemisorption behaviour we have studied on these alloys.

2. GLASSY TRANSITION METAL ALLOYS

The valence band spectra of three different compositions of glassy Pd-Zr alloys are shown in Fig.2 along with the spectra of the pure polycrystalline alloy constituents [9]. The alloy spectrum is dominated by a distinct two peak structure which does not exhibit any similarities with the spectra of Pd and Zr. Moreover it is obvious from Fig.2 that the alloy spectrum cannot be obtained by a superposition of the pure element valence band spectra. From this we can already conclude that Pd-Zr is a strongly interacting system for which no simple picture such as a rigid band model can be applied. This behaviour is in contrast with weakly interacting systems such as the random compositionally disordered binary alloys with alloy constituents close to each other in the periodic table which exhibit quite often a rigid band behaviour [5].

An important question in connection with the alloy valence band spectrum in Fig.2 concerns the local electronic structure or, in other terms, what is the contribution from the Pd 4d and the Zr 4d-electron states to the dominating peaks at E_F and at a binding energy of about 3 eV? A first qualitative information to this problem can be obtained by looking at a difference spectrum obtained from measurements on samples with two different alloy cocentrations. From this it can clearly be concluded that an increase in the Pd content leads to an increase in the intensity of the peak at 3 eV and correspondingly a decrease in the intensity for the peak at E_F (see also Fig.2).

This qualitative picture can be confirmed by analizing other experimental data with more local character such as X-ray emission spectroscopy (XES), Auger electron spectroscopy (AES) and XPS core level line shape analysis [5]. Figure 3 shows the X-ray emission spectra of the Pd and Zr emission bands from the alloy $Pd_{30}Zr_{70}$ [11]. Since the recombination of the electron from the valence band to the L_{III} shell is subject to dipole selection rules ($\Delta l = \pm 1$) the emission spectra are determined essentially by transitions of Pd and Zr 4d-electrons. Since XES measures basically the local d band density of states with respect to a core level (Pd $2p_{3/2}$ and Zr $2p_{3/2}$, respectively) the XES spectra in Fig.3 have been shifted by the $2p_{3/2}$ binding energies as determined by XPS (from the corresponding $3p_{3/2}$ binding energies) in order to locate the position of the Fermi level. The XES emission bands clearly show the local nature of the two peaks present in the photoemission valence band spectra: the peak at the higher binding energy is mainly related to transitions at the Pd site whereas the peak close to E_F is due to Zr 4d electron states. In addition, the spectra indicate a distinct contribution from the Zr 4d states at a binding energy where the "Pd peak" is located, clearly revealing the mixing between Pd and Zr 4d electron states due to hybridization [11]. Figure 4 shows a comparison of the UPS valence band spectrum of glassy $Pd_{25}Zr_{75}$ with calculated density of states [10]. The calculation is performed by the self-consistent ASW (augmented spherical wave) method for the fcc-like $AuCu_3$ type symmetry. The calculated state densities shown here refer to the theoretical equilibrium lattice separation obtained by energy minimization. The only input to these calculations are the atomic numbers of the constituents and the crystal structure. The site decomposed densities of states are included in addition to the total density of states. The comparison of the experimental and theoretical DOS shows a close agreement between corresponding d-band positions. In addition, the calculations clearly confirm the important features obtained experimentally: the dominance of the d-states of the early transition metal at E_F and the position of the Pd 4d-states at a binding energy of about 3 eV.

The d-band behaviour of glassy Pd-Zr alloys as discussed above has basically been observed in all glassy alloys with late and early transition metals (i.e. transition metals with more than half and less than half filled d-bands, respectively). However, distinct trends in the d-band shifts on alloying with respect to the relative position of the two alloy constituents in the periodic table and the alloy concentration has been found. Figure 5 shows a comparison of different valence band spectra of glassy Zr-alloys measured with 21.2 eV photons [15). An analysis similar to that described above for Pd-Zr revealed for all the alloys shown in Fig.8 a dominance of the Zr 4d-electron states at E_F and, again, the d-bands related to the late transition metals are shifted to higher binding energies with respect to the pure metals. As can be seen from the spectra shown in Fig.5a the d-band splitting depends on the valence difference (or in other terms on the group number difference) of the two alloy constituents: the increase in the group number difference is correlated with an increase in the d-band splitting. Figure 5b demonstrates the influence of a change of the late transition metal within the same group in the periodic table. The separation between the peaks related to the d-bands of the two alloy constituents is abviously increasing if we replace Ni by the heavier elements Pd and Pt.

An interesting observation can be made if we try to correlate the d-band binding energy shifts ΔE_B of the late transition metal on alloying and the glass forming ability. In fact it has been shown that the glass forming ability is increasing within the sequence Fe-Zr, Co-Zr, Ni-Zr and Cu-Zr [16]. Since the d-band binding energy shifts are correlated with the alloy heats of formation we can conclude that the systems with high d-d interaction (i.e. stronger A-B bonds) exhibit an increased glass forming ability among the alloys which do form glasses.

As already mentioned in the introduction the knowledge of the electronic structure of these alloys is important at least for the qualitative understanding of many physical properties. As an example, the electrical resistivity depends strongly on the angular momentum projected density of states at the Fermi level and a classification of alloy resistivities depending on whether an open d-band at E_F is present or not can be made. In addition, since the electronic structure at E_F is essentially determined by the early transition metal and the contribution from this at E_F is about the same in alloys between a given early transition metal and different late transition metals we find about the same resistivities for the alloys in such a group. Also some trends in the superconducting transition temperature T_c can be understood in terms of the valence band structure. This has been discussed in more detail elsewhere [5)].

3. ACTINIDE GLASSES

The possibility to change the alloy concentration continuously over a fairly large range is an important advantage in studies using metallic glasses. This has been applied in the context of uranium glasses in order to increase the mean U-U distance in an amorphous matrix and to examine the electronic configuration of uranium. A study of U-rich binary glasses with Fe, Co, and Ni did reveal a U $5f^3$ configuration [17)]. Therefore, no change in the valence between these alloys and pure alpha-U has been found. However, the situation can be different in a dilute U-glass as we can see in Fig.6 [18)]. The UPS valence band spectra of the glassy, alloys $(Pd_{25}U_{75})_{82}Si_{18}$, $(Pd_{95}U_5)_{83}Si_{17}$ and $Pd_{82}Si_{18}$ and pure uranium are shown in Fig. 6a. The U-rich alloy is dominated by a U 5f peak at E_F and shows only a weak Pd 4d peak at a binding energy of about 4 eV. The corresponding U 4f core level spectrum is shown in Fig.9b. The close similarity to the U 4f spectrum of pure U is evident: the core level binding energy shift is quite small (0.2 eV) and the shape of the core lines exhibit almost no change on alloying which indicates that no change in the electronic configuration takes place. A drastic change, however, can be observed if we lower the U content. The U peak at E_F is no longer discernible in the valence band spectrum (even at very high magnification) and the U 4f core levels exhibit a marked shift of 1.4 eV and a distinct change in the shape of the core line accompanied with the occurence of a satellite. An almost identical behaviour has been found in crystalline Pd_3U in which the U is considered to be in the tetravalent U $5f^2$ configuration [19)]. Therefore, uranium exhibits a change in the valence as a function of alloy concentration. This observation has been confirmed by magnetic measurements which indicate a large increase in the magnetic moment per U atom in the dilute U alloy compared with the U-rich system [20]. Measurements of U glasses in other amorphous matrices such as Ni-Y-U or Ni-Zr-U behave differently and no change in the valence could be detected [18].

4. CHEMISORPTION OF CARBON MONOXIDE ON GLASSY TRANSITION METAL ALLOYS

Some amorphous alloys exhibit unusual surface properties such as high corrosion resistivity or high catalytic activity [21]. Despite these facts only few investigations using surface sensitive analytical methods have been used so far in order to understand the surface reactions and the role of substrate involved in corrosion and catalysis. As a first model reaction the chemisorption behaviour of CO on glassy transition metal alloys has been examined and an interesting correlation between the dissociation behaviour and the local electronic structure has been found.

A typical chemisorption experiment with CO on glassy $Ni_{91}Zr_9$ is shown in Fig. 7 and the reaction of pure Zr and Ni with CO is included for comparison [22,23]. The qualitatively different behaviour of Ni and Zr is evident from Fig.7. Pure polycrystalline Ni exhibits molecular adsorption at room temperature which is typical for the late transition metals whereas Zr, like most other early transition metals, dissociates the adsorbed CO molecule. This can be seen unambiguously from the peaks related to the 4-sigma and 5-sigma(1-pi molecular orbitals at about 8 and 11 eV in the case of Ni and the O 2p and C 2p peak at about 3.5 and 6 eV in the case of Zr. The spectrum of the CO exposed $Ni_{91}Zr_9$ alloy shows clearly a mixture of peaks which can be attributed to both molecular and dissociative adsorption. A quantitative analysis yields about an equal amount of dissociative and molecular adsorption at room temperature for this alloy.

The temperature dependence of the dissociation properties are of particular interest. The amount of molecular and dissociative adsorption for Ni and glassy $Ni_{91}Zr_9$ is shown as a function of temperature in Fig.8 [22]. The measurements on the glassy alloy reveal a complete change from molecular to dissociative adsorption in the temperature range between -100^oC and $+100^oC$, whereas only the desorption of molecularly adsorbed CO has been observed on polycristalline Ni. An increase of the Zr content in the glassy alloy leads to a further decrease of the temperature at which the change from molecular to dissociative adsorption occurs. From this it is obvious that the adsorption behaviour of these alloys cannot be understood in terms of Ni-like and Zr-like adsorption sites but the local electronic structure, which is very different from that in the corresponding pure metals, has to be taken into account.

The split d-bands of the transition metal alloys under investigation with well separated contributions from the two alloy constituents make it possible to identify the electron states which are responsible for dissociation and molecular adsorption. In the Ni-Zr alloys the Ni d-band reveals an increasing binding energy with decreasing Ni content with respect to pure Ni (see also Fig.5). Since the same binding energy shifts have been observed for the 4-sigma and 5-sigma/1-pi molecular orbitals we conclude that the molecular adsorption is essentially coupled with the antibounding Ni 3d-electron states. On the other hand the relevance of the bonding Zr 4d-electron states at E_F for the dissociation of CO can be seen from Fig.9. The $Ni_{37}Zr_{63}$ alloy has been exposed to CO at room temperature and 100% dissociative adsorption has been found [22]. Comparing the broken curve (after dissociation) with the spectrum of the clean alloy surface shows a distinct decrease in intensity at E_F, while the intensity of the Ni 3d peak at about 2 eV remains practically the same in both spectra.

The above observations are not specific for Ni-Zr alloys but are also found in other glassy transition metal alloys. In all the cases studied so far the bonding d-states at E_F of an early transition metal seems to be crucial for the dissociation of the CO molecule. As soon as these electron states are missing at E_F (replaced by p-electron states), dissociation is no longer observed e.g. in Pd-Si, Ni-Al, Fe-Ni-B or similar alloys. In these alloys a decrease in the CO-substrate bond occurs, visible from a reduced temperature for CO desorption. However, the position and symmetry of the different electron states seem to be responsible for the CO adsorption behaviour, i.e. changes in the C-O and the CO-substrate bond. On the other hand the CO adsorption behaviour (molecular, dissociative or desorption) can be controlled by the choice of the alloy constituents.

5. CONCLUSIONS

Glassy transition metal alloys exhibit strong d-d interactions which are manifested by a distinct d-band splitting. The degree of the splitting depends on the relative position of the alloy constituents in the periodic table and the alloy concentration. The strong d-d interaction is also responsible for the distinct change in the local electronic structure on alloying. The chemisorption behaviour of the glassy transition metal alloys is strongly determined by these local properties and the distinct separation of the d-states arising from the individual alloy constituents makes it possible to discriminate the role of the different d-electron states for molecular and dissociative adsorption.

ACKNOWLEDGEMENT

I would like to acknowledge fruitful collaborations and discussions in the field presented here with K.H. Bennemann, U. Gubler, H.-J. Güntherodt, C.F. Hague, R. Hauert, P. Häussler, G. Indlekofer, V.L. Moruzzi, R. Schlögl, D. Tomanek and A.R. Williams. This work has been supported by the Swiss National Science Foundation and the "Kommission zur Förderung der wissenschaftlichen Forschung".

REFERENCES

1) see e.g. "Photoemission in Solids I", eds. M.Cardona and L. Ley, Topics in Applied Physics, Vol. 26, Springer-Verlag, Berlin, Heidelberg, New York 1978
"Low Energy Electron and Surface Chemistry", eds. G. Ertl and J. Küppers, Verlag Chemie, Weinheim 1974.

2) Gubler, U.M., Indlekofer, G., Oelhafen, P., Güntherodt, H.-J. and Moruzzi, V.L., Proc. V Int.Conf. on Rapidly Quenched Metals, ed. S. Steeb and H. Warlimont, Elsevier Science Publishers B.V., p.971, 1985.

3) Yeh, J.J. and Lindau, I., Atomic Data and Nuclear Data Tables 32, 1 (1985).

4) Nagel, S.R. and Tauc, J., Phys. Rev. Lett. 35, 380 (1975).

5) Oelhafen, P., in "Glassy Metals II" eds. H. Beck and H.-J. Güntherodt, Topics in Applied Physics, Vol. 53, Springer-Verlag, Berlin Heidelberg, 1983, p.283.

6) Häussler, P., Baumann, F., Krieg, J., Indlekofer, G., Oelhafen, P. and Güntherodt, H.-J., Phys. Rev. Lett. 51, 714 (1983).
Häussler, P., Baumann, F., Gubler, U., Oelhafen, P. and Güntherodt, H.-J., Proc. V Int.Conf. on Rapidly Quenched Metals, eds. S. Steeb and H. Warlimont, Elsevier Science Publishers B.V., 1985, p.1007.

7) Nagel, S.R., Gubler, U.M., Hague, C.F., Krieg, J., Lapka, R., Oelhafen, P., Güntherodt, H.-J., Evers, J., Weiss, A., Moruzzi, V.L. and Williams, A.R., Phys. Rev. Lett. 49, 575 (1982).

8) Oelhafen, P., Krieg, J. and Güntherodt, H.-J., unpublished results.

9) Oelhafen, P., Hauser, E., Güntherodt, H.-J. and Bennemann, K.H., Phys. Rev. Lett. 43, 1134 (1979).

10) Moruzzi, V.L., Oelhafen, P., Williams, A.R., Lapka, R., Güntherodt, H.-J. and Kübler, J., Phys. Rev. B27, 2049 (1983).

11) Hague, C.F., Fairlie, R.H., Gyorffy, B.L., Oelhafen, P. and Güntherodt, H.-J., J. Phys. F. (Metal Phys.), 11, L95 (1981).

12) Oelhafen, P., Hauser, E. and Güntherodt, H.-J. in "Inner Shell and X-ray Physics of Atoms and Solids" eds. D.J. Fabian, H. Kleinpoppen and L.M. Watson, Plenum Press, New York 1981, p.575

13) Fuggle, J.C., in "Electron Spectroscopy: Theory, Techniques and Applications", Vol. 4 ed. by C.R. Brundle and A.D. Baker, Academic Press, London p.85, 1981.

14) Oelhafen, P. unpublished results.

15) Oelhafen, P., Hauser, E. and Güntherodt, H.-J., Solid State Commun. 35, 1017 (1980).

16) Nishi, Y., Morahoshi, T., Kawakami, M., Suzuki, K. and Masumoto, T. in "Proc. 4th Intern. Conf. on Rapidly Quenched Metals", eds. T. Masumoto and K. Suzuki", p.111, 1982.

17) Oelhafen, P., Indlekofer, G., Krieg, J., Lapka, R., Gubler, U.M., Güntherodt, H.-J., Hague, C.F. and Mariot, J.M., J. Noncryst. Sol. 61 62, 1067 (1984).

18) Indlekofer, G., Oelhafen, P. and Güntherodt, H.-J., Proc. V Int. Conf. on Rapidly Quenched Metals, eds. S. Steeb and H. Warlimont, Elsevier Sciences Publishers B.V., p.1011, 1985.

19) Baer, Y., Ott, H.R. and Andres, K., Solid State Commun. 36, 387 (1980).

20) Durand, J. et al., to be published.

21) Schlögl, R., Proc. V int.Conf. on Rapidly Quenched Metals, eds. S. Steeb and H. Warlimont, Elsevier Science Publishers B.V., p.1723, 1985.

22) Hauert, R., Oelhafen, P., Schlögl, R. and Güntherodt, H.-J., Solid State Commun., 55, 583 (1985).

23) Tomanek, D., Hauert, R., Oelhafen, P., Schlögl, R. and Güntherodt, H.-J., Surf. Sci. 160, L493 (1985).

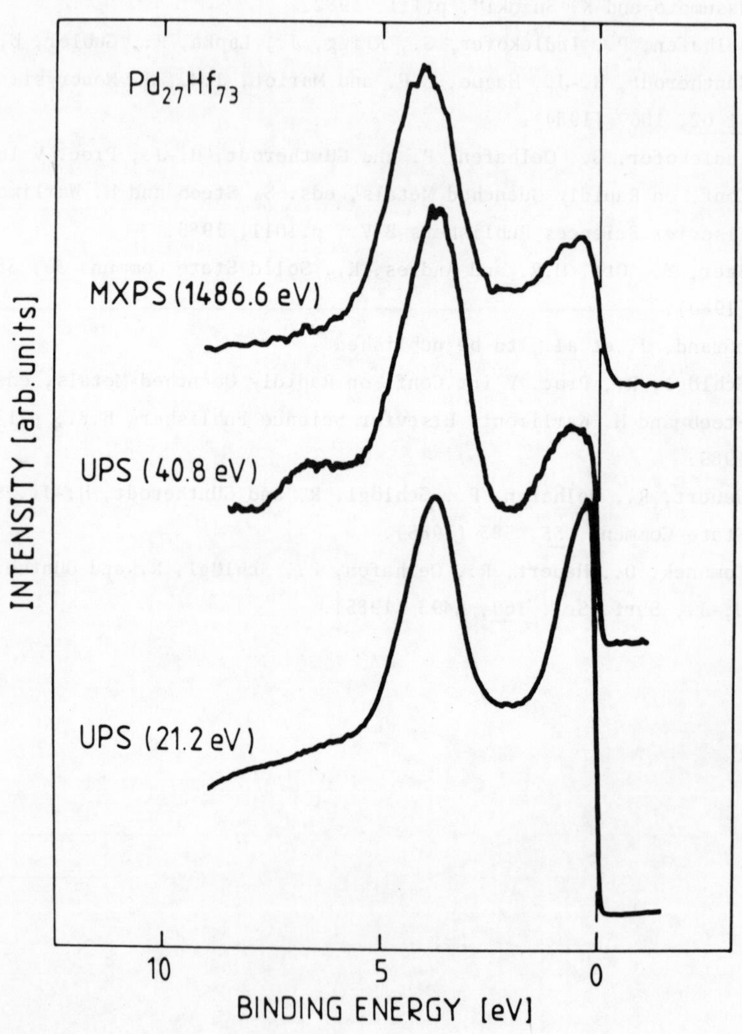

Fig.1. Photoelectron valence band spectra of glassy $Pd_{27}Hf_{73}$ measured with different excitation energies [2)]

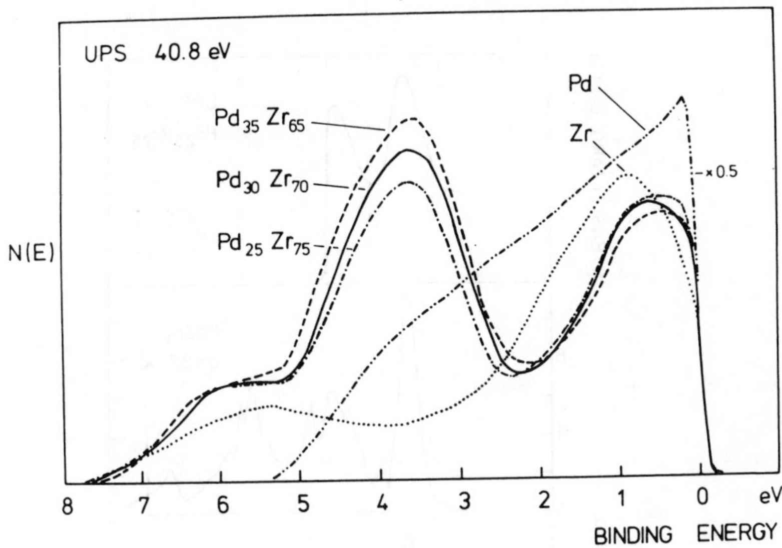

Fig.2. UPS valence band spectra of glassy Pd-Zr alloys and polycrystalline Pd and Zr [9)]

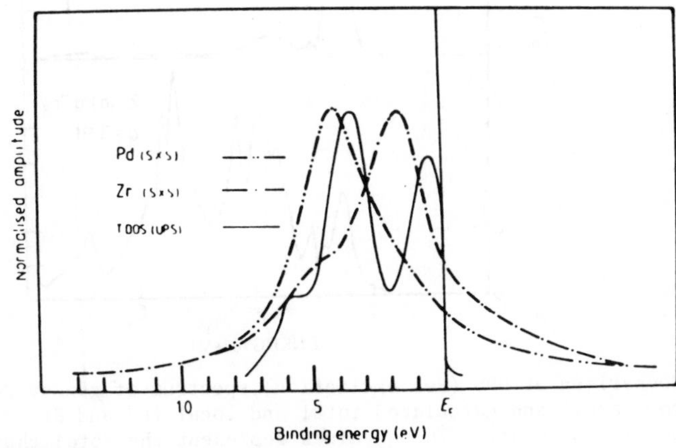

Fig.3. Local distributions of Pd and Zr according to XES measurements. The emission bands have been positioned with respect to E_F by means of XPS core level determinations [11)]

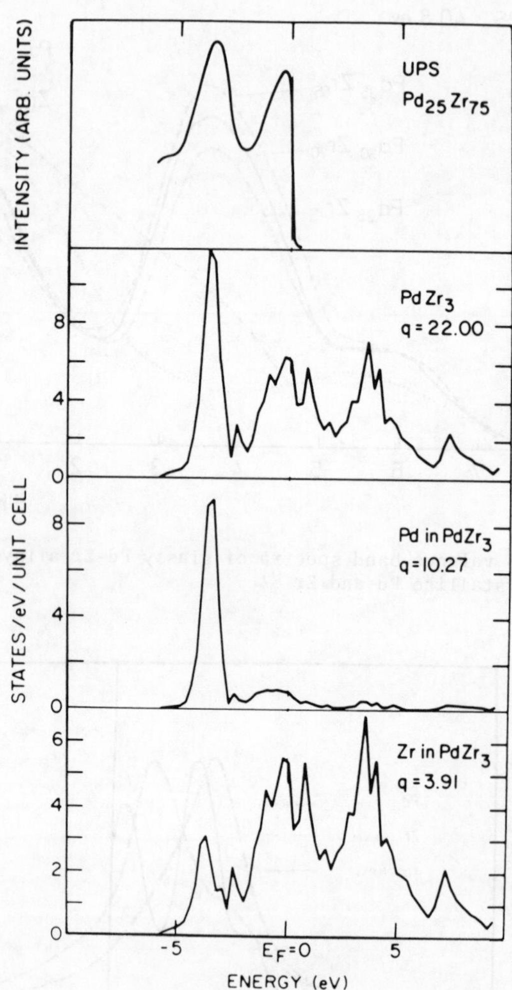

Fig.4. Comparison of UPS (HeI excitation) spectrum of glassy $Pd_{25}Zr_{75}$ top panel) and calculated total and local (Pd and Zr site) density of states. The q-values represent the total charge per unit cell and the local charge at the Pd and Zr site, respectively [10].

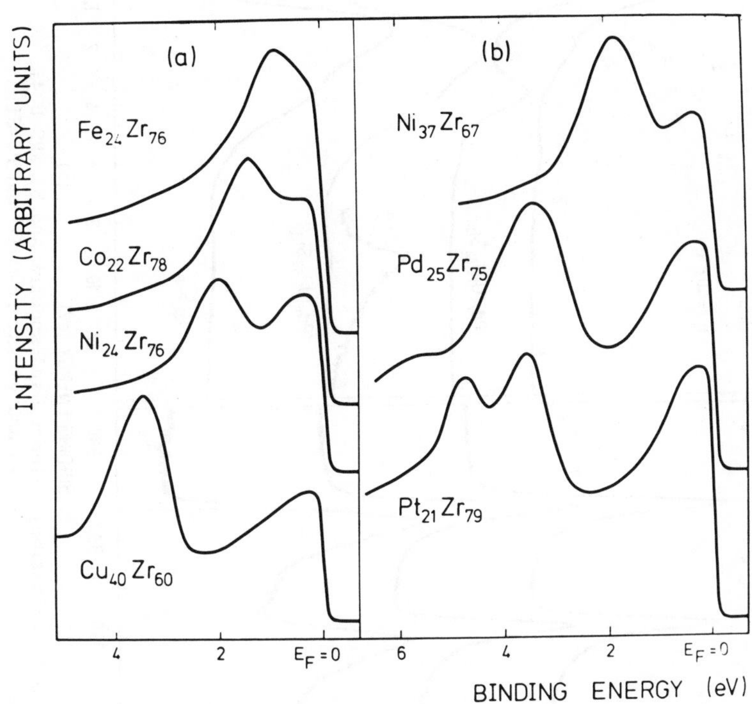

Fig.5. UPS (HeI excitation) spectra of glassy Zr alloys with late transition metals from the same series (a) and from the same group of the periodic table (b) [15]

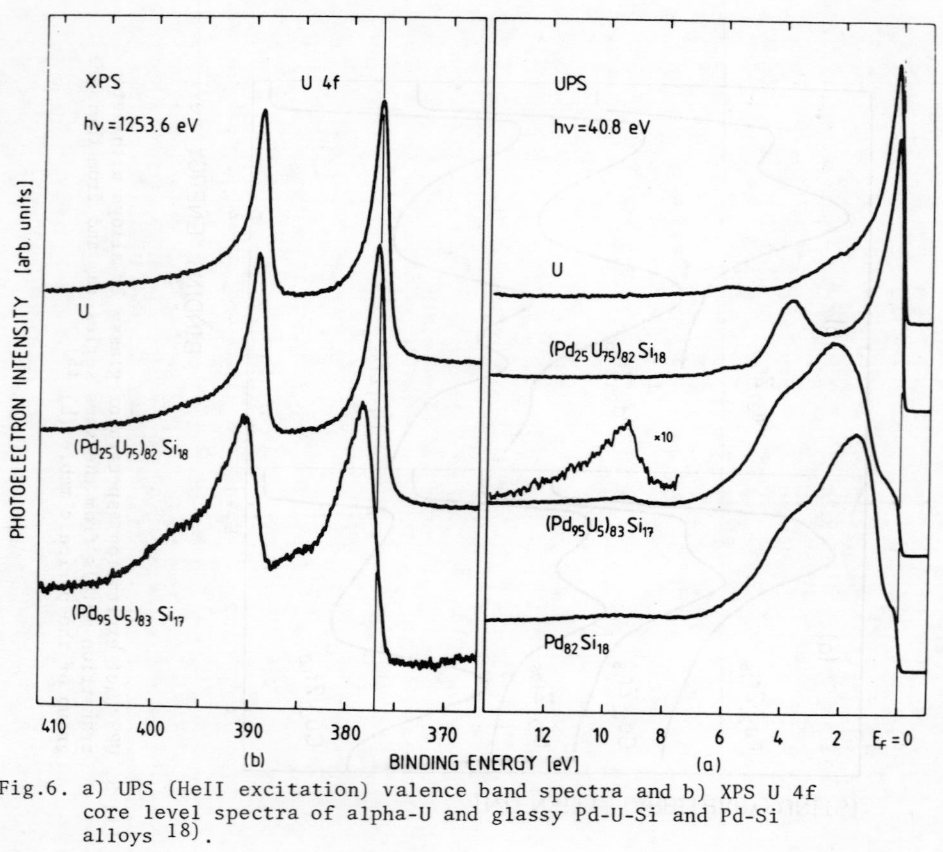

Fig.6. a) UPS (HeII excitation) valence band spectra and b) XPS U 4f core level spectra of alpha-U and glassy Pd-U-Si and Pd-Si alloys 18).

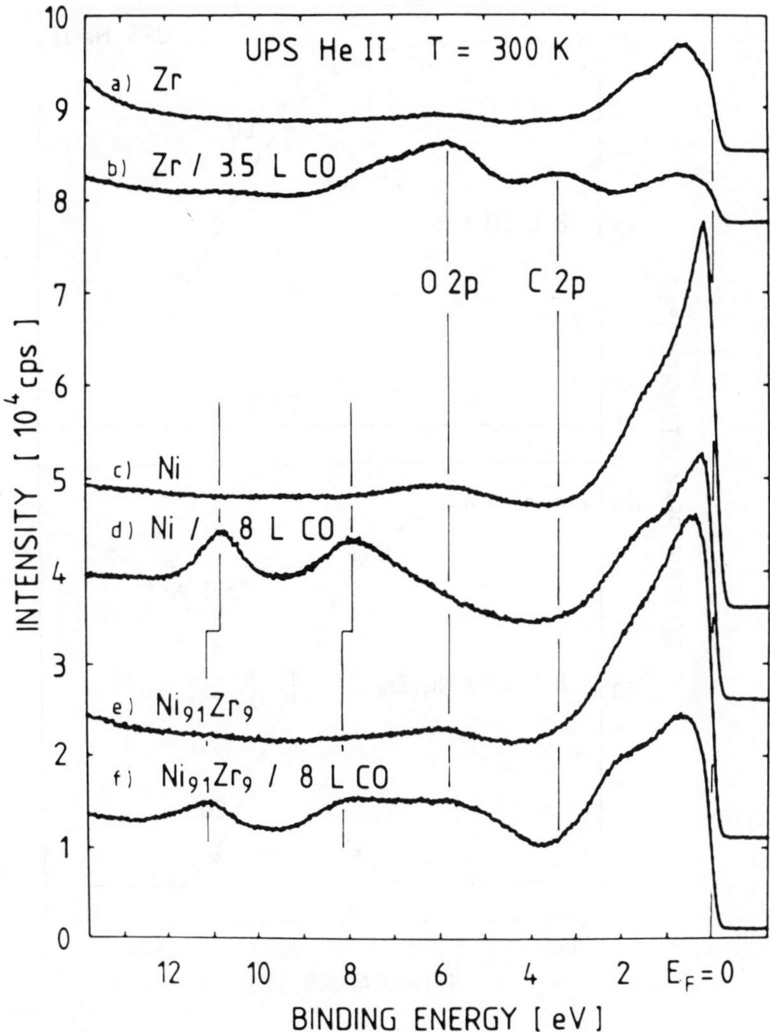

Fig.7. UPS valence band spectra of clean polycrystalline Zr (a), Ni (c), glassy $Ni_{91}Zr_9$ (e) and the corresponding surfaces after CO exposure (curves b, d and f) [22]. 1 L (1 Langmuir) corresponds to an exposition of $1 \cdot 10^{-6}$Torr·sec.

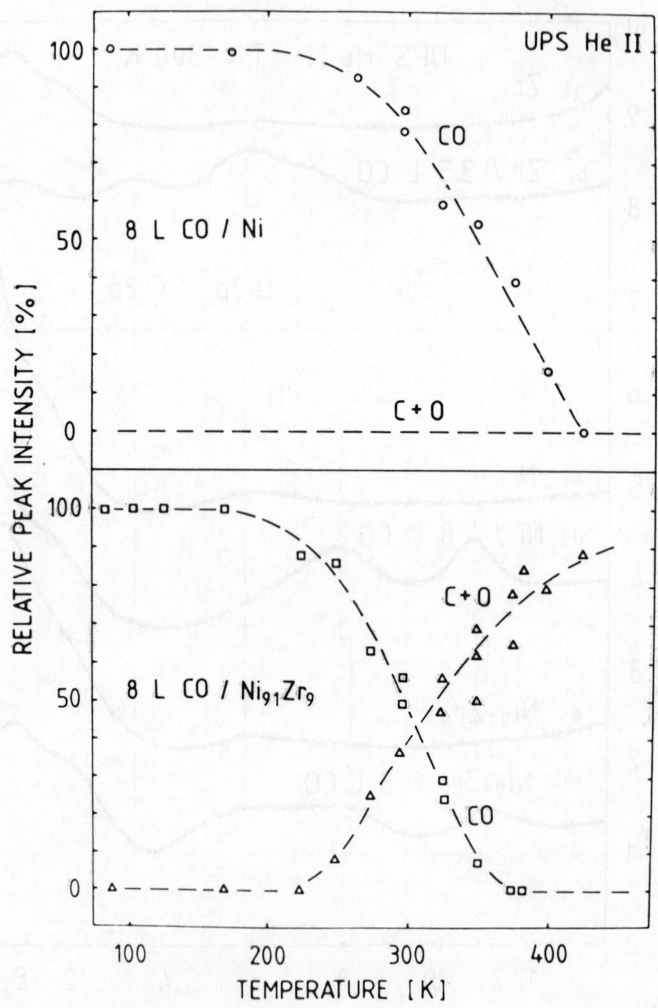

Fig.8. Amount of molecular and dissociative CO adsorption as a function of temperature for amorphous $Ni_{91}Zr_9$ (lower panel) and polycrystalline Ni (upper panel). In the latter case no dissociation is observed and the curve labeled "CO" shows the desorption behaviour as a function of temperature [22].

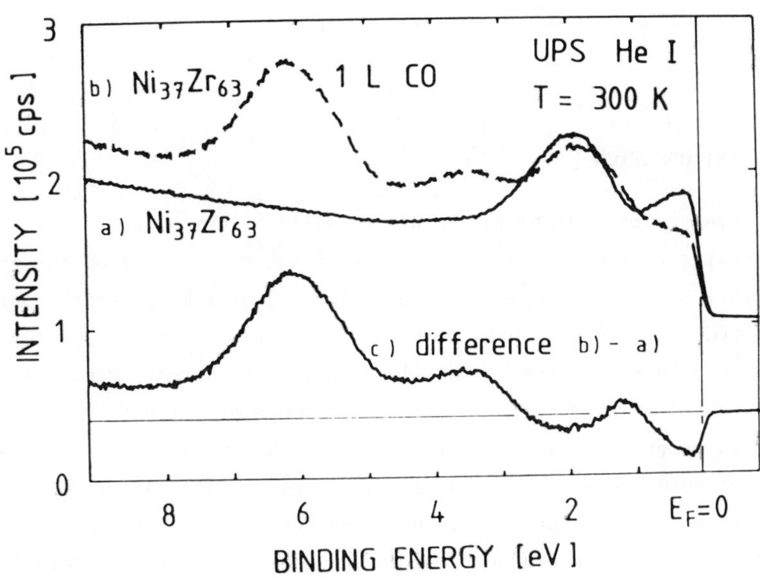

Fig.9. UPS valence band spectra of clean glassy $Ni_{37}Zr_{63}$ (a) and after an exposition of 1 Langmuir CO (b) and difference spectrum (c) [22].

APPLICATIONS OF NMR WITH EMPHASIS ON AMORPHOUS OR DISORDERED ALLOYS

Jacques Durand
Physique du Solide (UA CNRS 155) Universite de Nancy I
France

1. INTRODUCTION

Even restricted to metals and metallic alloys or compounds, a comprehensive review of the applications of NMR would overflow the space limitations of this paper. In the following, will be presented only few selected examples as representative of historical advances in the physics of metallic systems or as illustrative of current trends in todays research. This overview is divided into two parts. The first one refers to paramagnetic metals and alloys, for which emphasis is placed on studies of quadrupole and dipolar interactions, joint analysis of Knight-shifts and relaxation rates, investigations of the particular case constituted by spin glasses. The second part is devoted to metallic ferromagnets.

2. PARAMAGNETIC METALS AND ALLOYS

2.1. Quadrupole and Dipolar Interactions

Although they are observable by NMR in many magnetically ordered systems, the quadrupole and dipolar interactions are more often studied and more readily analysed in paramagnetic metals and alloys. A few illustrations are presented concerning basic physics such as Friedel charge oscillations and mixed-valent Rare-Earth compounds or dealing with atomic arrangements in quasicrystals and metallic glasses or atomic motion such as hydrogen diffusion in crystalline metals and non-crystalline alloys.

2.1.1. Charge oscillations around impurities in a metal. Knowledge of electric field gradients (EFG) as can be achieved by nuclear techniques is of twofold interest in metallic systems since they arise both from the charge distribution in the lattice and from the distribution of the conduction electrons. In crystalline alloys, especially when the symmetry of the matrix is cubic, quadrupole effects can be studied as a measure of the aspherical charge density of conduction electrons around an atomic site [1]. Numerous NMR studies along this line have been performed for various impurities in Cu and in Al [2,3], providing one of the most straightforward demonstrations of long-range screening charge oscillations around impurities in a metal.

2.1.2. Quadrupole interactions in mixed-valent Rare-Earth compounds. Rare-Earth compounds with non-integer valence are a new challenging field of research where NMR studies of quadrupole interactions have proven informative. In such compounds, the homogeneous admixture of valence for the Rare-Earth element is reflected in the one hand by an anomaly in the lattice parameter and in the other hand by some unusual physical properties originally attributed to fast fluctuations between two electronic configurations differing in the occupation number of the 4f electronic shell. Obviously, both lattice and conduction electrons will contribute to the electric field gradient parameters in a non-trivial manner, making conjectural the determination of the motionally averaged valence from the measurement of the EFG parameters at a fixed temperature in a non-cubic mixed-valent compound. But relative variations of the valence as a function of thermodynamic parameters such as temperature or pressure are expected to be more readily detectable through EFG parameters.

This has been done by comparing the temperature dependence of the quadrupole interactions by NMR on ^{63}Cu in the three $SmCu_2Si_2$,$YbCu_2Si_2$

and Eu Cu_2 Si_2 compounds of same crystal structure [4]. The first compound is "normal" (Sm is trivalent). The quadrupole interaction follows then the expected $T^{3/2}$ law. In the second compound, Yb is mixed--valent, but with a temperature independent admixture of valence. In this case, the temperature dependence of the EFG parameter seems to be regular (fig.1). The mixed-valent character in this case can be extracted only by the magnitude of the quadrupole interaction, which is not an easy task. In the latter compound, the anomalous temperature dependence of the quadrupole interaction reflects the temperature dependence of the Eu valence.

2.1.3. <u>EFG parameters in non-crystalline alloys</u>. In non-crystalline alloys, the study of EFG parameters yields some information about the local symmetry around a given atomic site. Such a piece of information cannot be readily obtained from standard spectroscopic techniques. Recent NMR work along this line has been carried out in quasi-crystals and in amorphous alloys. The recent discovery of long-range orientational order with icosahedral point symmetry in rapidly solidified Al based alloys has prompted a large amount of experimental work and computer simulations in an effort to determine the actual atomic positions in these alloys. Among other site selective techniques, NMR was used to investigate the local symmetry in Al_{86} Mn_{14} quasi-crystalline alloy [5]. ^{27}Al and ^{55}Mn spin-echo spectra were compared for this alloy in the amorphous and quasi-crystalline phases and in the crystalline equilibrium state. No discernible quadrupole structure was observed for the Mn sites in both phases. Quadrupole effects were detected at the Al sites and analysed by using an average value of the quadrupole frequency $\bar{\nu}_Q$, a range of values for the asymmetry parameter η and a Gaussian distribution of ν_Q with second moment $\Delta\nu_Q/\bar{\nu}_Q$. Experimental spectra were satisfactorily reproduced with second moment smaller in the icosahedral than in the amorphous phase (0.3 and 0.4, respectively) and with a smaller range of η values in the quasi-crystal ($0.1 \leqq \eta \leqq 0.6$) than in the amorphous alloy ($0.1 \leqq \eta \leqq 0.9$), while the same value of $\bar{\nu}_Q$ was used in both cases. From this rather preliminary study, local symmetries appear to be fairly similar at both atomic sites in the amorphous and icosahedral modifications of this Al based alloy. Further investigations in other quasi-crystalline materials along with more refined analysis of the EFG parameters are clearly needed in this new growing field.

Due to its relative seniority, the field of amorphous alloys has been more widely investigated than that of quasi-crystals. Consequently, more achievement has been obtained by NMR in the knowledge of the average symmetry about the resonant atomic sites in different amorphous alloys. NMR has proven a powerful technique for that purpose, especially when the atomic probe has a low nuclear spin (I = 3/2) and is the minority element ("glass former") of the amorphous alloy [6]. NMR data together with results of electric-field-gradient studies by Mössbauer spectroscopy have yielded an increasing amount of evidence that the Dense-Random-Packing Model of Hard Spheres represents an adequate zero--order approximation to describe the atomic scale structure of amorphous alloys when the short-range order is loosely defined. But this model fails to describe the variety of local symmetries experimentally evidenced in metallic glasses [7] (fig.2). On the other hand, there is no simple one-to-one correspondence between the local symmetry about a constituent in an amorphous alloy and in the crystalline compound of same composition. This point was clearly illustrated [8] by NMR on ^{11}B in a series of amorphous $Ni_{1-x}B_x$ alloys (fig.3) and in related crystalline compounds (fig.4). Both similarities (in Ni_3B, for example), and discrepancies (in Ni_2B) of the local symmetry around B in the amorphous and crystalline phases can be explained in the following way. Some elementary motifs (such as trigonal prisms) can be easily accommodated in a non-periodic packing. Other motifs (such as Archimaedian antiprisms) seem to be practically excluded in an amorphous packing. Such information beyond the pair correlation functions are unique, providing a way to discriminate between the various structural models compatible with the pair correlation functions yielded by standard diffraction techniques applied to metallic glasses. The potential of NMR studies of the effects of thermal history on glass structure is well documented for non-metallic glasses [9]. Such a field has remained practically unexplored for metallic glasses so far.

2.1.4. Quadrupolar and dipolar interactions in metal-hydrogen systems. Crystalline metal-hydrogen systems constitute an area of earliest and broadest application of NMR [10]. In particular, diffusion constant of hydrogen can be directly measured using NMR. If the atomic structure of the host metal is known, diffusion mechanisms can be identified. If the host is an amorphous alloy, one needs a comparison with hydrogen diffusion data in a crystalline compound of same composition. Deuterons with spin I = 1 have a quadrupole moment. Significant information about structural effects have thus been obtained through deuteron quadrupole interaction in crystalline V, Nb and Ta. For protons which do not possess a nuclear quadrupole moment, dipolar interactions are the leading mechanism governing the proton NMR linewidth, spin-lattice and rotating-frame relaxation times. Such measurements as a function of temperature over the thermally activated regime allow an estimation of the activation energies involved and of their distribution. From proton NMR investigations performed in crystalline and amorphous Zr-Pd and Ti-Cu hydrides [11], the following conclusions were obtained. First, a non-Arrhenius behaviour for the thermally activated diffusion seems to be the rule in amorphous hydrides. Second, a comparison of hydrogen diffusion in amorphous and crystalline compounds of same composition must be handled with care, since opposite conclusions might be drawn depending on the hydrogen content in the amorphous host. Indeed, in amorphous alloys with low hydrogen content, hydrogen atoms can be trapped in free volume zones (analogous to defects or voids) and therefore the diffusion can be found slower than in crystalline counter-parts. When the hydrogen content is high, hydrogen diffusion is generally enhanced in amorphous alloys, the structural disorder possibly providing some diffusion paths with higher activation energies than in the related crystalline compounds.

2.2. Knight-shift and Relaxation Rate.

Combined studies of the Knight-shift and of the spin-lattice relaxation time T_1 are a powerful tool to get an estimation of the contributions to the local susceptibility. After a short survey of such studies in paramagnetic metals and alloys, we summarize the applications of NMR to some more specific problems, namely the mechanisms of superconductivity, the characterization of metallic multilayers and catalysts and the onset of a magnetic moment.

2.2.1. Contributions to local susceptibility of metals and alloys. Analysis of Knight-shift and T_1 have constituted an important historical step in our basic understanding of magnetism in pure transition metals. By neglecting exchange effects and electron correlations, and with reasonable estimations or calculations of the local fields, the different components of the susceptibility, namely the contact, core polarization and Van Vleck contributions, were determined for a large number of transition metals, and their systematic variation along a transitional series was interpreted in terms of partial densities of states [12,13].

In binary, ternary alloys, such a quantitative analysis is not a trivial one for most cases. At least, the experimental value of the Korringa ratio for a given atomic species is always very informative. Indeed, from comparison with theoretical values of this Korringa ratio for s, p or d electrons, one gets an insight about the predominent symmetry of the electrons at the Fermi level, and thus about the degree of hydridization or charge transfer. Such a knowledge in metallic glasses might shed some light on the atomic structure and on the cohesion in these alloys [14÷16].

2.2.2. NMR in superconductors. NMR is known to have played an historical role in both verifying and questioning the Bardeen-Cooper-Schrieffer (BCS) theory of superconductivity [17]. In BCS superconductors, all electrons occupy pair states at zero temperature. Raising the temperature results in breaking up some Cooper pairs to form single-particle excitations which contribute to nuclear relaxation. At very low temperature ($T << T_c$), the relaxation process is thermally activated. A direct measure of the superconducting gap can be obtained from $T_1^{-1} \alpha \exp(-\Delta/k_BT)$ where Δ is the gap energy. At temperatures just below T_c, a peak in T_1^{-1} (T) has been observed in many "conventional" superconductors and attributed to the opening out of the gap in the excitation spectrum. The experimental value obtained for the gap can be compared with the BCS value ($2\Delta/k_B T_c = 3.52$) calculated in the limit of weak electron-phonon coupling. In order to reach the intrinsic superconducting properties and to avoid sophistications due to the mixed state in type II superconductors, one might use field cycling techniques or nuclear quadrupole resonance (NQR) in zero static field. Most recently, the temperature dependence of T_1 has proven a very successful tool to probe the superconducting state in new-explored superconductors. We refer to two examples of these "exotic" superconductors, namely, the amorphous superconductors and the heavy-fermion superconductors.

The first amorphous superconductors to be studied were simple metals or simple metal based alloys. It has been argued that the structural disorder in these superconductors might result in a softening of the phonon spectrum and thus in a considerable enhancement of the electron-phonon coupling [18]. The question has then arisen whether this enhancement occurs also for transition metal based amorphous superconductors. An answer to this question was given by ^{31}P NMR measurements on an amorphous $(Mo_{0.5}Ru_{0.5})_{80}P_{20}$ alloy [19]. Measurements were performed in applied field low enough with respect to critical field Hc_2 for the results to be considered as characteristic of the zero--field state. T_1 (T) was found to follow the behaviour predicted by theory for normal type I superconductors, namely the feature just below T_c related to the gap opening (fig.5) and the low-temperature exponential behaviour (fig.6) yielding for the gap Δ a value practically identical to that of the BCS theory in the weak coupling limit. It was then shown, in agreement with tunnelling and specific heat measurements, that the enhancement of electron-phonon coupling is not a general rule in transition-metal based amorphous superconductors.

Very recently, an intensive interest has developed in superconducting 4f ($CeCu_2Si_2$) and 5f (U based) compounds, where the very large values of the low temperature specific heat coefficients in the normal state suggest highly enhanced conduction-band effective masses. Hence, these compounds are commonly labelled as heavy fermion systems. An enormous specific heat jump was observed at T_c in these superconductors, indicating that the heavy quasi-particles are involved in the

Cooper pairing. The question is open as to the pairing mechanism whether it is a singlet pairing with an isotropic energy gap or a triplet pairing. Spin-lattice relaxation measurements on ^{9}Be in U Be_{13} [20] and on ^{63}Cu in $CeCu_2Si_2$ [21] have confirmed the unusual nature of superconductivity in these compounds as suggested by specific heat and ultrasonic attenuation measurements. Indeed, the feature observed near T_c in BCS superconductors is vanishingly small in the T_1 (T) of these heavy fermion superconductors. On the other hand, $(T_1)^{-1}$ varies as T^3 for $T << T_c$ for both compounds (fig. 7), with, however, a deviation from this power law below 0.2 K in U Be_{13}. Such a behaviour is consistent with anisotropic pairing models for which the gap vanishes along lines on the Fermi surface. This anisotropic energy gap can occur either in a singlet pairing hypothesis or in a polar-state model for L = 1 triplet pairing. However, in this latter hypothesis, the agreement of experimental data with model calculations would be qualitative only. More data at lower temperature are clearly needed to judge on this controversial point.

2.2.3. Characterization of surfaces and interfaces in metallic multilayers and in catalysts. Metallic superlattices represent a new class of artificial materials which, from an ideal point of view, can be designed as to exhibit a variety of physical properties not usually encountered in natural materials. These metallic superlattices can be classified in two groups. In one case, one expects the interdiffusion at interfaces to remain extremely low, since the properties of interest rely on the periodic stacking of bilayers. In the other case, the interdiffusion is expected to be fast enough for a metastable or stable alloy or compound to be formed, or, in kinetically and thermodynamically favourable diffusion couples, for an amorphous alloy to be obtained.

For this latter family of multilayers, the metallic superlattice is the initial stage that allows a rigourous control of the composition of the final product. In both types of metallic superlattices, the characterization of the interfaces is a crucial, but not trivial task. NMR measurements could play a key role along this line as can be shown through two recent examples. Cu NMR experiments were performed on a Cu-Nb superlattice of 7.69 μm total thickness, with 5.75 nm thickness for individual bilayers, meaning on average 13.8 atomic planes in each Cu layer [22]. From these pulsed NMR studies along with T_1 measurements Cu resonances were identified as "bulk-Cu" an "shifted-Cu" resonances, respectively. Small changes in the isotropic Knight-shift in "bulk-Cu" signal as compared with pure metallic Cu were observed together with the onset of anisotropy on the Knight-shift. On the other hand, the lower value of the Korringa term $T_1 T$ suggested an enhancement of the Cu density of states. The "shifted-Cu" signal was attributed to 4 or five atomic layers on both sides of the central region ("bulk-Cu"). This interfacial region is characterized by large isotropic and anisotropic frequency shifts along with very low density of electronic states.

Another recent NMR study was devoted to the characterization of ^{51}V at the interface of a Fe-V superlattice [23]. Due to the presence of magnetic Fe layers, part of the V nuclei are submitted to a transferred hyperfine field, which is broadely distributed. The remaining V atoms have a "non-magnetic" behaviour and their Korringa relaxation rate was experimentally determined. Analysis of the hyperfine field distribution on magnetic V nuclei and of the static and dynamic data on "non-magnetic" V nuclei led the authors to the following conclusions. The interface atomic layer would contain an admixture of Fe and V atoms in the proportion 50% - 50%. The two layers next to the interface atomic layer would contain about 5% counter elements. However, from the anisotropic Knight-shift and the reduction of the density of states at the Fermi level observed on "non-magnetic" V sites it is concluded that the influence of the Fe layers is extended somehow over the interior regions in the V layers.

NMR study of Pt catalysts in the form of small metallic particles supported on alumina presents, in common with the above-mentioned studies, the fact of dealing with a large fraction of atoms on the surface [24]. Platinum metal is a typical catalyst. Heterogeneous catalysis is a subject of great interest from the joined points of view of technological applications and of basic science. The catalytic reaction takes place at the surface of the platinum particles. Relatively little is known about processes at surfaces in general, and hence about microscopic mechanisms governing heterogeneous catalysis. A systematic ^{195}Pt NMR study of Pt catalysts including analyses of line shapes, Knight--shifts, spin-lattice and spin-spin relaxation times was carried out with the aim of better characterize Pt atoms in the bulk and on the surface of the catalytic particles [24]. The following conclusions were obtained. First, Pt particles even as large as 100 Å in diameter do not exhibit properties of bulk Pt. Second, Pt atoms on the surface of the particles give rise to a distinct peak (a "surface peak") in the NMR adsorption line shape. These surface atoms are bonded to adsorbed molecules. Their Knight-shift is so vanishingly small that these surface atoms can be regarded as non metallic. The position and shape of the surface peak were found to dopend on the type of adsorbed molecule, hence providing a way to identify the adsorbed molecules. Along the same line, NMR was recently used [25] to study the motion of isolated carbon atoms on Pt clusters and to measure the diffusion energy from motional narrowing of ^{13}C line shape and from motion-induced ^{13}C spin--lattice relaxation. The carbon atoms were found to be very mobile with a low activation energy for translational motion.

2.2.4. The onset of a magnetic moment. Due to the presence of broad conduction bands in the matrix, atoms with partially filled d-shells, when diluted in a metallic host or when forming metallic compounds, do not generally carry in their ground state the value of magnetic moment predicted by the Hund rule. The problem of local moment formation in metallic systems was approached in different ways by both theoreticians and experimentalists [13]. Either, for a series of (diluted) impurities in a given host, one determines whether they are magnetic or non-magnetic at 0 K in this matrix. The onset of a localized moment is then relied to a criterion which involves the density of d-states at the Fermi level along with effective values on intra-atomic exchange and Coulomb interactions. Or, for a fixed host, one investigates the concentration range of a given impurity for localized moments to appear. The occurrence of magnetism will be then commonly discussed in terms of environment effect or impurity-impurity interactions. Or, for a given amount of impurity in a matrix, one studies the temperature range where the localized moment sets in. In this latter case, the properties of local moments are most frequently treated in relation with a characteristic temperature (Kondo or spin-fluctuation temperatures). NMR has been widely used in all these aspects of the problem of local moment formation either in impurity studies of the local susceptibility or in measurements of the polarization resulting in the host. We will refer in the following to a few examples of these complementary approaches.

Attention to the importance of local environmental effects in the onset of the magnetic moment was first drawn by JACCARINO and WALKER who measured the ^{59}Co shift and intensity in crystalline $Co_{0.01}(Rh_yPd_{1-y})_{0.99}$ alloys and deduced that a minimum of two palladium atoms nearest neighbours to a given Co site was required for Co to carry a moment [26]. Other examples of discontinuous moment formation were investigated by ^{59}Co NMR in binary or pseudo-binary alloys such as Cu-Co, Mo-Co and $Co_{0.01}(Ti_yMo_{1-y})_{0.99}$ alloys [13]. Many hyperfine studies were performed on impurities in the dilute limit in simple metal hosts or in transition metals. Spin and orbital contributions to the impurity susceptibility were successfully separated and the Kondo or spin-fluctuation effects were clearly identified.

Host hyperfine studies allow one to determine the impurity-induced spin polarization in the host. In solid alloys, the near-neighbour polarization results in a line broadening or in discrete satellites of magnetic origin in favourable cases, while the long range of the host spin polarization in exchange-enhanced matrixes often results in impurity-induced shifts of the host NMR line. In contrast, in liquid alloys, due to the rapid atomic diffusion in the liquid state, the host NMR is characterized by large resonance shifts and very little broadening which are thought to reflect the near-neighbour polarization. Thus, the magnitude of the NMR shift, as illustrated in fig. 8, clearly reveals the expected difference between 3d impurities in aluminium and copper.

To some extent, the problem of onset of magnetism in Ce or Yb compounds with sp elements presents some similarities with that of local moment formation on 3d impurities in a metal. Because of this analogy, these compounds are termed Kondo lattices, with a characteristic temperature (T_K) for the magnetic-non-magnetic cross-over. At lower temperature, the properties of such systems might indicate strong correlations of interaction Fermi liquids. Most of the NMR experiments in these compounds have been performed on non-lanthanide nuclei, but the coupling to 4f moments via indirect interactions is thought to be strong enough for the transferred hyperfine fields to reflect the 4f magnetic behaviour. A recent representative example is a measurement of the temperature dependence of T_1^{-1} on ^{63}Cu in $CeCu_6$, a Kondo lattice [27]. For T > 6 K (Kondo temperature), T_1^{-1} was found to exhibit a Curie-Weiss behaviour. T_1^{-1} is temperature dependent below 6 K to reach a Korringa behaviour (T_1 T = constant) below 0.2 K, temperature at which a Fermi-liquid state sets in. Along the same line, one should refer to Ti NMR study of the nearly ferromagnetic Fermi liquid $TiBe_2$[28]. In particular, NMR has showed that the M/H anomaly observed at low temperature in such a system characterized by a spin-fluctuation temperature of about 20 K is mainly induced by homogenous modifications of the electronic structure.

2.3. NMR in Spin Glasses

A single impurity in a metal with infinite electronic mean-free--path creates in its neighbourhood a conduction-electron spin density proportional to cos 2 k_F r/r^3 in its asymptotic form (k_F being the Fermi momentum and r the distance away from the impurity). A second impurity will interact with the first one via this polarization of the conduction electrons. Due to its oscillatory nature, this interaction (Ruderman-Kittel-Kasuya-Yoshida or RKKY interaction) between two impurities will be positive or negative depending on the distance r. For moderately concentrated magnetic alloys, this admixture of interactions of opposite sign can give rise to an original type of short-range magnetic order called spin glass, to mean that below a characteristic temperature the spins are frozen-in like atoms in a glassy structure. The spin-freezing temperature T_G is defined by a sharp cusp in zero--field susceptibility. Below T_G, field-cooling effects and time relaxation phenomena are observed. Experimentalists and theoreticians have accumulated an over-whelming amount of works showing the complex nature of the spin-glass ordering process [29]. We quote a few examples of the NMR contribution to that field of research.

A first important question was addressed about the influence of shortening the electron mean-free-path on the spin-density oscillations in a host metal. The temperature dependence of the ^{63}Cu line width in a series of $\underline{Cu}_{0.95-x}Mn_{0.05}.Al_x$ alloys indicated a strong reduction in polarization with decreasing mean-free-path (increasing Al concentration)[30]. The RKKY interaction was found to be damped as a function of the mean-free-path λ by a factor exp $(-r_o/\lambda)$ where r_o is a characteristic distance determined from experience. As for the spin-glass freezing process, its great sensitivity to applied field imposes for NMR studies the use of zero or small external field. Among the most informative host nuclear resonance experiments in spin glasses, let us quote the line-width ^{63}Cu NMR measurements in Cu-Mn spin glasses with Mn content within the range of 1 at.% [31] giving evidence for a frozen configuration of Mn spins below T_G. From T_1 measurements on ^{63}Cu in very dilute Cu-Mn spin glasses with 12 to 43 ppm Mn [32] it was concluded that the Mn spins freeze gradually below a well-defined temperature T_G at which spin motions are highly correlated. Small applied field was used in these two latter sets of experiments.

The magnetic state of a Cu-Mn (1 at.%) spin glass was studied in the zero-field cooled (ZFC) and in the field-cooled states (FC) through zero-field Cu NMR on the first (resonance A) and fourth (resonance B) nearest neighbours of Mn atoms [33]. Main results for resonance B are summarized in fig. 9. The enhancement factor η of the rf field seen by the nuclei was determined for different sample conditions, namely in the ZFC cooled state with different values of a H_i applied field and also on the sample cooled under different applied fields H_c. This enhancement factor is attributed to domain rotation induced by the rf field, as in ferromagnetic materials (see section 3 above). In the ZFC sample, when no field is applied ($H_i = 0$), the enhancement factor is practically zero implying the absence of independent domains. The domain structure is induced by the magnetic field applied on the ZFC sample (H_i) or during the cooling process (H_c). The field dependence of η in fig. 9 resembles very much the field dependence commonly observed for the isothermal (H_i) and thermal (H_c) remanent magnetization in spin glasses below T_G. These induced domains are not simple ferromagnetic clusters. From the scaling of η with $I_{SC}^{1/2}$ (fig. 9), I_{SC} being the signal detected in a single coil, it is concluded that most of the Mn spins are involved in these domains.

Finally, let us mention a very recent zero-field NMR study of ^{55}Mn in Cu-Mn, Ag-Mn and Au-Mn spin glasses [34]. Collective excitations in these spin glasses were analysed through the temperature dependence of the Mn local magnetization. A T^2 dependence of the relative frequency shift of the NMR line was observed for $T < 0.2\ T_G$ as expected for magnons with a linear dispersion relation and a large stiffness constant.

3. FERROMAGNETIC METALS AND ALLOYS

This survey of applications of NMR to magnetically ordered metallic systems will be restricted to ferromagnetic materials. We first illustrate briefly the special features of NMR in metallic ferromagnets which are related namely to domains and walls, to demagnetizing fields and to the existence of a spontaneous hyperfine field. Then, we give a few examples of the information that can be gained by NMR on the physics of ferromagnets and in physical metallurgy. Finally, we refer to recent use of NMR in the study of ferromagnetic superconductors.

3.1. Special Features of NMR in Ferromagnets

3.1.1. Domains and walls. A ferromagnet is characterized by the existence of a spontaneous electronic magnetization and, hence, an hyperfine field H_F which allows in principle the observation of a NMR signal in zero external field. On the other hand, the bulk of a ferromagnet is spontaneously divided into domains separated by walls. As mentionned above about spin glasses, nuclei in domain walls experience a rf field which is considerably enhanced in good ferromagnets so that one observes principally those nuclei in a zero-field NMR experiment done at low rf power. This was the case in the first zero field NMR experiment in pure cobalt [35]. The question then arose as to whether the resonance frequencies were the same for nuclei in domains and in domain wall centers. This frequency shift was found to be very weak (0.5 MHz) in fcc Co [30] and well-defined in hcp Co [37], where, as illustrated in fig. 10 for spectra taken at 290 K, the 214 MHz peak is attributed to nuclei at domains wall edges, while the 221 MHz peak comes from nuclei at the center of domain walls. This was accounted for by the anisotropy of the hyperfine field, and by the fact that the easy axes of magnetization are such as to favour different orientations, with respect to the crystallographic axes, of the electron spins in domains and domain wall centres.

3.1.2. <u>NMR at zero-field and under external field</u>. When an external field is applied on a multidomain ferromagnetic sample, the field dependence of the resonance frequency is typically as illustrated in fig. 11 [38]. As long as the technical saturation is not reached, i.e. for fields smaller than $H_L = H_a + H_D$ (where H_a is the anisotropy field and $H_D = NM$, the demagnetizing field of the sample), the walls take a new equilibrium state so that H_L and H balance each other. As a result, the signal amplitude strongly decreases with H, while the field dependence of the resonance frequency is weak or practically zero. For $H > H_L$ the frequency varies linearly with external field H, according to:

$$\nu(H) = \pm \left(\frac{\gamma}{2\pi} + K\right) \left[H_F + H_L + H\right] ,$$

where K is the Knight-shift, the positive and negative signs indicating that H_F is positive or negative, respectively. In transition metals (Co, Fe, Ni), H_a is small with respect to H_D, so that H_L can be taken for H_D.

NMR under applied field is then an absolute requirement if one is interested in intrinsic values of longitudinal and transverse relaxation times, since for $H < H_L$ the times T_1 and T_2 would be strongly affected by wall motions in multidomains samples. On the other hand, the field dependence of resonant frequency for $H > H_L$ allows one to determine the sign of H_F and the value of the Knight shift as will be discussed later. In addition, from the value of the applied field H at which $\nu(H)$ starts to vary linearly with H and also from a comparison between the experimental zero-field frequency $\nu(H)$ and the frequency obtained from a zero-field extrapolation of the linear dependent $\nu(H)$, one is able to obtain valuable information about the geometrical shape of the sample. Such a piece of information has been systematically analysed for single-domain Co particles [39)] and for the first crystallization products of a thermally treated amorphous Co-P alloy [40)].

3.1.3. NMR in Transition Metal and in Rare-Earth ferromagnets. In Transition Metals, due to the quenching of orbital moment, the magnetization is basically related to electronic spins. On the other hand, the shape of the NMR spectrum will reflect the local environment in the form of line broadening or/and of discrete satellites. Examples of the information yielded by these lineshapes will be given below. In Rare Earths, the electronic magnetization arises mainly from the 4f shells which have a small spatial extension and are practically shielded by the more extended s and d states. On the other hand, in Rare-Earth ions with $L \neq 0$, the line shapes will reflect basically the orbital contributions to H_F [41)].

3.2. Typical Information Obtained by NMR in Transition Metal Based Ferromagnets.

The few selected applications of NMR in metallic ferromagnets that we present in the following illustrate different aspects of the information accessible through NMR experiments: the resonance frequency and its field dependence, the static structures of the NMR line and the relaxation of the nuclear spins.

3.2.1. Knight-shift in 3d ferromagnets: Fe, Co, Ni. As already shown in fig. 11, for bcc Fe, the field derivative of the NMR line for $H > H_L$ in the ferromagnetic state may exhibit some departure from the expected gradient, i.e. from nuclear value of $\gamma/2\pi$, making possible an experimental determination of the Knight shift in ferromagnets, $K = (\gamma_{eff} - \gamma_n)/\gamma_n$. Values of K obtained this way are 90 (± 27), 194 (± 25) and 78 ± 10 (all in 10^{-4}) for Ni [42], hcp Co [43] and bcc Fe [38], respectively. From the Knight-shifts, values of the Pauli spin paramagnetism susceptibility $\chi_P = \chi_P^d + \chi_P^s$ were determined (in 10^{-6} emu $mole^{-1}$) as 44, 71 and 168 (± 25), for Ni, Co and Fe, respectively, while the Van-Vleck susceptibility was found to be 76, 202 and 143 (± 25), all in units of 10^{-6} emu $mole^{-1}$, for Ni, Co and Fe, respectively. These values provide a direct confirmation of band models for ferromagnetism in 3d metals. They constitute a critical test for band structure calculations.

3.2.2. Impurity hyperfine fields in 3d ferromagnets. The exact determination of the different contributions to the hyperfine field in a ferromagnet is a very difficult problem [44]. Impurity hyperfine fields in a ferromagnetic host is a more tractable problem, especially for s-p impurities [45]. Among other nuclear techniques, NMR has brought an important contribution in the knowledge of the HF systematics for series of impurities in 3d ferromagnets [46]. An example of this systematics is given in fig. 12. A change of sign for H_F occurs in each series of s-p impurities (3 sp, 4 sp, 5 sp, 6 sp): H_F goes from negative to positive values when increasing the atomic number in each series. Such a behaviour has been qualitatively explained by a screening model for non-magnetic impurities in a band ferromagnet [45].

A direct approach of the impurity density of states in ferromagnets is provided by the Korringa rate $(T_1\ T)^{-1}$ obtained by spin-lattice relaxation time measurements on the impurity NMR signal. Measurements have been performed under fields larger than H_D in order to get rid of the wall contributions to the spin relaxation. This has been successfully done for 3d elements diluted in Ni, Co,[47] and Fe[48], clearly showing the occurrence of virtual bound states at the Fermi level.

3.2.3. Selective site substitution of transition metal impurities in the Fe_3 Si matrix. Fe_3 Si has a fcc crystal structure of the DO_3-type, with four sites denoted A, B, C and D. The Fe atoms occupy two different sites [A,C] and [B] which are chemically and magnetically inequivalent. The Fe [A,C] have 4 Fe [B] and 4 Si as first-near neighbours (1 nn) and magnetic moment of 1.35 μ_B, while the Fe [B] have 8 Fe [A,C] 1 nn and carry a moment of 2.2 μ_B. Through NMR measurements (confirmed later on by neutron diffraction and Mössbauer spectroscopy results), it has been shown that transition metal impurities selectively substitute for Fe in one of the two inequivalent sites [49]; namely the 3d elements to the left of Fe such as Mn and V, show a strong preference for the [B] sites, while those to the right of Fe such as Co, Ni, select the [A,C] sites. This site selectivity was deduced for the two above-mentioned cases from spin-echo spectra taken on ^{59}Co (fig. 13) and ^{55}Mn (fig. 14) diluted in stoichiometric and offstoichiometric Fe_3Si. Frequencies of the NMR peaks are found to depend essentially on the number of Fe 1 nn. For the Co substituted Fe_3 Si, the single line for the stoichiometric compound corresponds to 4 Fe and 4 Si 1 nn, while the three lines for the offstoichiometric case are attributed to a co-ordination number for the first surrounding atomic shell of 8 atoms made of 4 Fe + 4 Si, or 5 Fe + 3 Si or 6 Fe + 2 Si. Such an atomic co-ordination characterizes the [A, C] sites. In contrast, as shown in fig. 14 for the Mn substitution, the Mn lines for both the stoichiometric and offstoichiometric cases are compatible in frequencies and relative intensities only with the Fe [B] sites whose environment is made of 8 Fe [A, C] 1 nn, 6 Si 2 nn and some combination of 12 Fe [B] and Mn 3 nn. This site selectivity was found to account for the values of local and bulk magnetization and also for the limits of solubility of transition metals in Fe_3 Si together with the stability of the DO_3 structure.

3.2.4. Interplay between magnetism and chemical short-range order. As in the previous study, the structure of the impurity NMR line was used in a Fe-2 at. % Co alloy to define the atomic arrangements and their evolution as a function of the annealing temperature T_a over a temperature range spanning the Curie temperature T_c [50]. A typical ^{59}Co

spin echo spectrum for $T_a = 700^oC$ is shown in fig. 15. From the concentration dependence of the relative intensity of the satellites in the $Fe_{1-x}Co_x$ alloys, satellites S_1, S_2 and S_3 were attributed to Co pairs while the lines S_3' and S_3'' were assigned to Co triplets. Only the Co pairs satellites were retained to study the two-site correlation functions expressed in terms of the Cowley SRO parameters α_1, α_2 and α_3. The evolution as a function of annealing temperature of α_i for the solution I where S_1 and S_2 were assigned to first and second n n pairs, respectively, is shown in fig. 16. The three SRO parameters undergo a marked change of slope around T_c, clearly indicating the coupling between magnetic energy and chemical SRO in ferromagnetic alloys in agreement with model predictions.

3.2.5. Coexistence of ferromagnetism and superconductivity [51]. The problem of coexistence of superconductivity and magnetic ordering in an homogeneous material has been recently renewed with the investigations on Chevrel phases such as $HoMo_6S_8$ and on Matthias phases such as $ErRh_4B_4$. As first suggested by Matthias and Suhl [52], the conditions for the occurrence of superconductivity in a ferromagnet are more favourable near a domain wall where magnetization is inhomogeneous and where electrons move in a field which varies in direction. Among other local techniques NMR has proven very efficient in checking the homogeneous nature of this coexistence since it enables to sense both phenomena at the same time, for example the ferromagnetism through the zero--field enhancement factor and the superconductivity through the temperature dependence of T_1. As an example, fig. 17 shows the temperature dependence of the spin-echo decay time T_2 of ^{157}Gd in $(Ce_{1-x}Gd_x)Ru_2$ [51]. For ^{157}Gd $(I = 3/2)$, $\frac{1}{T_2} \simeq 4/T_1$; the decay time reflects basically the spin-lattice relaxation T_1. For the x = 0.12 sample, below $T_c = 4.4$ K, one first observed $T_2 T$ = constant. Then below 0.1 K, the spin-lattice relaxation time exhibits a departure from linear behaviour versus T, departure characteristic of the superconducting state. Analysis of the data leads to the conclusion that in zero external field the sample is in a self-induced vortex state analogous to the nearly gapless state near Hc_2 in a type II superconductor.

CONCLUSION

Less sensitive than the Mössbauer spectroscopy, less versatile also for high temperature measurements, the NMR spectroscopy has become of moderate use during the last ten years for the study of metallic systems, in spite of the good resolution which characterizes this technique. Interest in NMR spectroscopy has become more intense among chemists, biophysicists, experts in organic metals or in low-dimensionality materials than among common physicists or physical metallurgists. This review was partially aimed at demonstrating that much can still be learned through NMR in the new fields of metallurgy or solid state physics such as metallic superlattices, quasicrystals, amorphous alloys, finely dispersed perticles, surfaces or strongly correlated electronic systems.

REFERENCES

1) Gautier, F., "Propriétés électroniques des métaux et alliages", eds. Janot, Chr. and Gerl, M. (Masson, Paris 1973), chap. IV. pp. 57-254 and chap. IX, by Minier, M. pp. 453-473.
2) See, for example, Minier, M. and Ho Dung, S., J. Phys. F, 7, 503-513 (1977); Berthier, C. and Minier, M., J. Phys. F, 7, 515-520 (1977).
3) For recent reviews, see Watson, R.E., Sternheimer, R.M. and Bennett, L.H., Phys. Rev. B, 30, 5209 (1984); Grüner, G. and Minier, M., Adv. Phys., 26, 231-284 (1977).
4) Sampathkumaran, E.V., Gupta, L.C. and Vijayaraghavan, R., Phys.Rev. Lett., 43, 1189-1192 (1979); id., in Valence Fluctuations in Solids, eds. Falikov, L.M., Hanke, W. and Maple, M.B. (North-Holland, Amsterdam 1981), pp. 241-244.
5) Warren, W.W., Jr., Chen, H.S. and Hauser, J.J., Phys. Rev. B, 32, 7614-7616 (1985).
6) See the reviews of Durand, J. in Glassy Metals II, ed. Beck, H. and Güntherodt, H.J. (Springer, Berlin 1983), Chapt. 11, 343-385; Panissod, P., Helvet. Phys. Acta, 58, 60-75 (1985).

7) Panissod, P., Aliaga Guerra, D., Amamou, A., Durand, J., Johnson, W.L., Carter, W.L. and Poon, S.J., Phys. Rev. Lett., 44, 1965-1967 (1980).

8. Panissod, P., Bakonyi, I. and Hasegawa, R., Phys. Rev. B, 28, 2374-2381 (1983).

9) Bray, P.J. and Holupka, E.J., J. Non-Cryst. Sol., 71, 411-428 (1985).

10) For a review on NMR in crystalline Metal-Hydrogen systems, see Cotts, R.M., in Hydrogen in Metals I, eds. Alefeld, G. and Völkl, J., Topics in Applied Physics 28 (Springer, Berlin 1978) chap. 9, pp. 227-265.

11) For a review, see Bowman, R.C., Jr., Cantrell, J.C., Attala, A., Etter, D.E., Craft, B.D., Wagner, J.E. and Johnson, W.L., J. Non-Cryst. Sol., 61-62, 649-655 (1984).

12) Jaccarino, V., in "Teoria del magnetismo nei metalli di transizione", Varenna School of Physics, 1966 (Academic, New-York, 1967) pp. 335-385.

13) Narath, A., CRC Crit. Rev. Solid State Phys., 3, 1-37 (1972).

14) Aliaga Guerra, A., Panissod, P. and Durand, J., Solid State Comm., 28, 745-749 (1978).

15) Hines, W.A., Glover, K., Clark, W.G., Kabacoff, L.T., Modelewski, C.U., Hasegawa, R. and Duwez, P., Phys. Rev. B, 21, 3771-3779 (1980).

16) Bennett, L.H., Schone, H.E. and Gustafsson, P., Phys. Rev. B, 18, 2027-2035 (1978).

17) For a review, see Mac Laughlin, D.E., Solid State Phys., 31, 1-69 (1976).

18) See, for example, Bergmann, G., Phys. Rep. C 27, 159-185 (1976).

19) Aliaga Guerra, D., Durand, J., Johnson, W.L. and Panissod, P., Solid State Commun., 31, 487-491 (1979).

20) Mac Laughlin, D.E., J. Mag. Mag. Mat., 47-48, 121-126 (1985); Mac Laughlin, D.E., Tien, C., Clark, W.G., Lan, M.D., Fisk, Z., Smith, J.L. and Ott, H.R., Phys. Rev. Lett., 53, 1833-1836 (1984).

21) Kitaoka, Y., Ueda, R., Kohara, T., Asayama, K., Onuki, Y. and Komatsubara, T., Proc. 5th Intern. Conf. Cryst. Field and Anomalous Mixing Effects in f. Electron Systems, April 1985 (to be published); id. Solid State Comun., 51, 461-466 (1984).

22) Yudkowsky, M., Halperin, W.P. and Schuller, I.K., Phys. Rev. B, 31, 1637-1639 (1985).

23) Takanashi, K., Yasuoka, H., Kawaguchi, K., Hosoito, N. and Shinjo, T., J. Phys. Soc. Japan, 53, 4315-4321 (1984).

24) Rhodes, H.E., Wang, P.K., Stokes, H.T., Slichter, C.P. and Sinfelt, J.H., Phys. Rev. B, 26, 3559-3568 (1982); Rhodes, H.E., Wang, P.K., Malowka, C.D., Rudaz, S.L., Stokes, H.T., Slichter, C.P. and Sinfelt, J.H., ibid. 3569-3574 (1982); Stokes, H.T., Rhodes, H.E., Wang, P.K., Slichter, C.P. and Sinfelt, J.H., ibid., 3575-3581 (1982).

25) Wang, P.K., Ansermet, J.P., Slichter, C.P. and Sinfelt, J.H., Phys. Rew. Lett., 55, 2731-2734 (1985).

26) Jaccarino, V. and Walker, L.R., Phys. Rev. Lett., 15, 258-261 (1965).

27) Kitaoka, Y., Fujiwara, K., Kohori, Y., Asayama, K., Onuki, Y. and Tomatsubara, T., J. Phys. Soc. Japan, 54, 3686-3689 (1985).

28) Alloul, H. and Mihaly, L., Phys. Rev. Lett., 48, 1420-1423 (1982).

29) For a recent review on spin glasses, see Fisher, K.H., Phys. Status Solidi (b), 116, 357-414 (1983); id. 130, 13-71 (1985).

30) Heeger, A.J., Klein, A.P. and Tu, P., Phys. Rev. Lett., 17, 803-805 (1966).

31) Levitt, D.A. and Walstedt, R.E., Phys. Rev. Lett., 38, 178-181 (1977); Mac Lauchlin, D.E. and Alloul, H., Phys. Rev. Lett., 38, 181-183, (1977).

32) Bloyet, D., Varoquaux, E., Vibet, C., Avenel, O. and Berglund, M.P. Phys. Rev. Lett., 40, 250-253 (1978).

33) Alloul, H., Phys. Rev. Lett., 42, 603-606 (1979).

34) Alloul, H. and Mendels, P., Phys. Rev. Lett., 54, 1313-1316 (1985).

35) Portis, A.M. and Gossard, A.C., J. Appl. Phys., 31, 205 S - 213 S (1960).

36) Tchao, Y.H. and Le Dang Khoi, C.R. Acad. Sci. (Paris), 260, 3886-3889 (1965).

37) Kawakami, M., Hihara, T., Koi, Y. and Wakiyama, T., J. Phys. Soc. Japan, 33, 1591-1598 (1972).

38) Oppelt, A., Kaplan, N., Fekete, D. and Kaplan, N., J. Mag. Mag. Mat., 15-18, 660-662 (1980).

39) Gossard, A.C., Portis, A.M., Rubinstein, M. and Lindquist, R.H., Phys. Rev. A, 138, 1415-1421 (1965).

40) This has been discussed at length by Durand, J., NMR applied to studies of metallic glasses, Atomic Energy Review, Suppl. no 1, 143-172 (1980).

41) For a review, see Barnes, R.G., in "Handbook of the Physics and Chemistry of Rare-Earths", eds. Gschneidner, K.A., Jr. and Eyring, L. (North-Holland, Amsterdam 1979), chap. 18, pp. 387-505.

42) Kropp, H., Dormann, E., Grayevsky, A. and Kaplan, N., Solid State Commun., 44, 1109-1112 (1982).

43) Fekete, D., Grayevsky, A., Shaltiel, D., Goebel, U., Dormann, E. and Kaplan, N., Phys. Rev. Lett., 36, 1566-1569 (1976).

44) For reviews and discussion of the problem, see "Magnetism", eds. Rado, G.T. and Suhl, H., vol. II A (Academic Press, New York 1965), especially, chap. 1, Freeman, A.J. and Watson, R.E., "Hyperfine Interactions", and chap. 6, Portis, A.M. and Lindquist, R.H., "Nuclear Resonance in Ferromagnetic Materials", pp. 357-383.

45) Daniel, E. and Friedel, J., J. Phys. Chem. Sol., 1601-1610 (1963).

46) Koster, T.A. and Shirley, D.A., Tables of Hyperfine Fields, in "Hyperfine Interactions in Excited Nuclei", eds. Goldring, G. and Kalish, R. (Gordon and Breach, New-York 1971), vol. IV, pp.1239-1254.

47) Durand, J., Gautier, F. and Robert, C., Solid State Commun., 11, 1213-1217 (1972).

48) Kontani, T., Hioki, T. and Masuda, Y., J. Phys. Soc. Japan, 32, 416-425 (1972).

49) Burch, T.J., Litrenta, T. and Budnick, J.I., Phys. Rev. Let.., 33, 421-424 (1974); Niculescu, V.A., Burch, T.J. and Budnick, J.I., J. Mag. Mag. Mat., 39, 223-267 (1983) and references therein.

50) Pierron-Bohnes, V., Cadeville, M.C. and Gautier, F., J. Phys. F, 13, 1689-1713 (1983).

51) Kitaoka, Y., Chang, N.S., Ebisu, T., Matsumura, M., Asayama, K. and Kumagai, K., J. Phys. Soc. Japan, 54, 1543-1551 (1985).

52) Matthias, B.T. and Suhl, H., Phys. Rev. Lett., 4, 51-54 (1960).

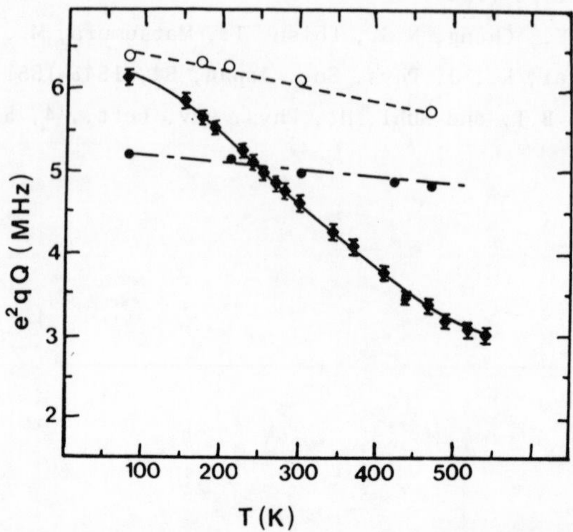

Fig.1. Dependence of the quadrupole interaction frequency (e^2 q Q) of ^{63}Cu on temperature for Sm Cu_2Si_2 (open circles), $YbCu_2Si_2$ (solid circles) and $EuCu_2Si_2$ (solid circles with error bars) (after ref.4).

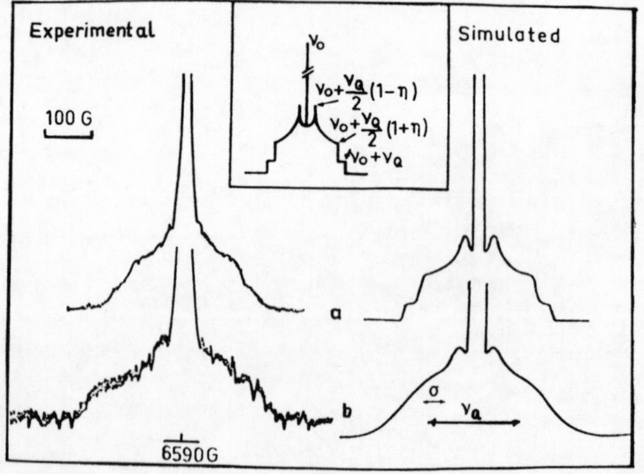

Fig.2. Experimental (T = 10 K) and simulated ^{11}B NMR spectra in c.Mo_2B (a); a.$Mo_{70}B_{30}$ (b) and a.$Mo_{48}Ru_{32}B_{20}$ (c). Inset: unbroadened "powder pattern" for $\nu_Q \neq 0$ and $\eta = 0$ (after ref.7).

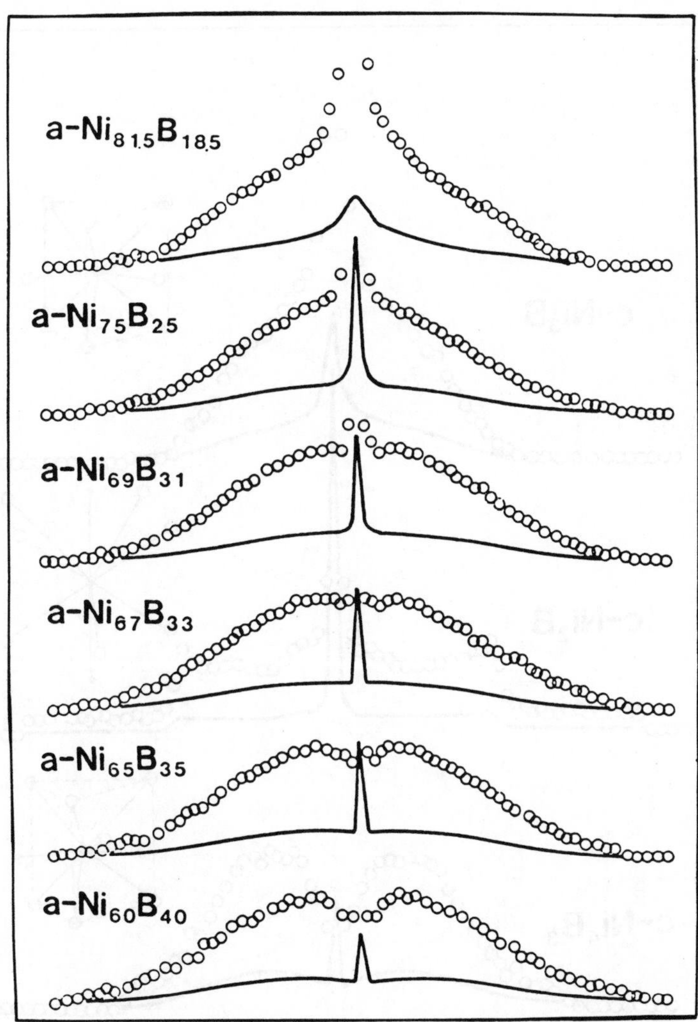

Fig.3. Experimental ^{11}B NMR spectra in amorphous NiB alloys for different compositions (Spectra were taken at 4.2 K, except for a-$Ni_{81.5}B_{18.5}$, T = 100 K) (after ref.8).

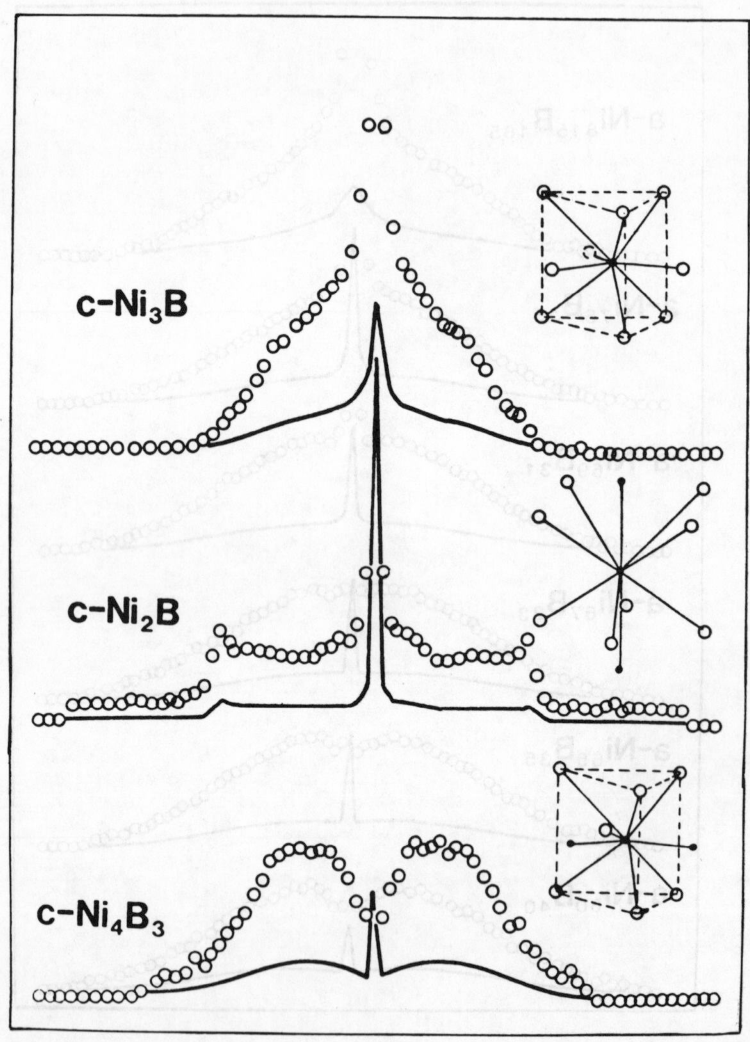

Fig.4. Experimental ^{11}B NMR spectra at 4.2 K in crystalline Ni_3B, Ni_2B and orthorhombic Ni_4B_3. Inset: boron co-ordination for each compound (after ref. 8).

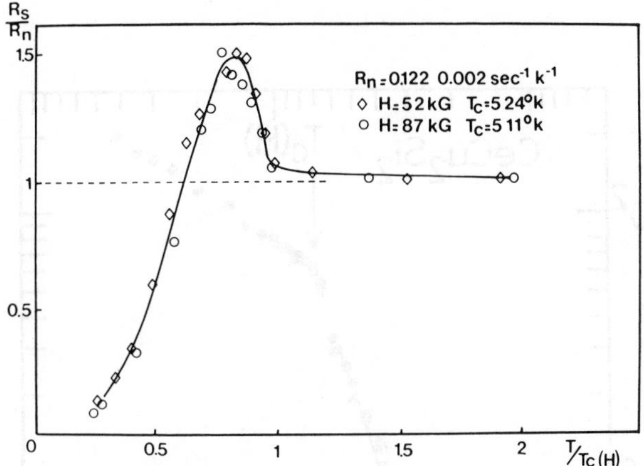

Fig.5. Reduced relaxation rate R_S/R_n versus reduced temperature T/T_c (H) (H = 5.2 and 8.7 kG) for ^{31}P in amorphous $(Mo_{0.5}Ru_{0.5})_{80}P_{20}$ (after ref. 19).

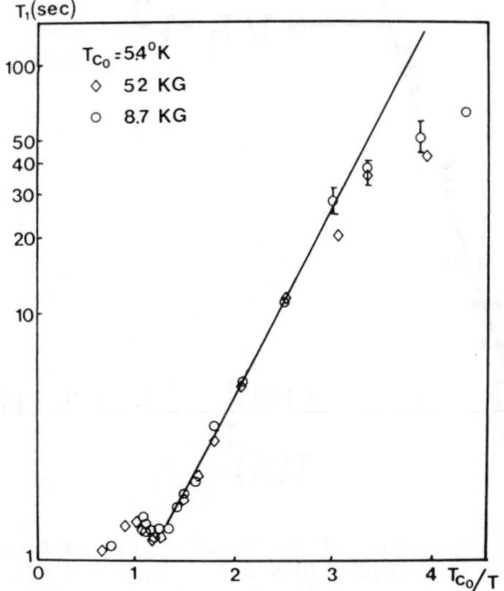

Fig.6. Spin-lattice relaxation time versus T_{c_0}/T (T_{c_0} = 5.4 K) for ^{31}P in amorphous $(Mo_{0.5}Ru_{0.5})_{80}P_{20}$.

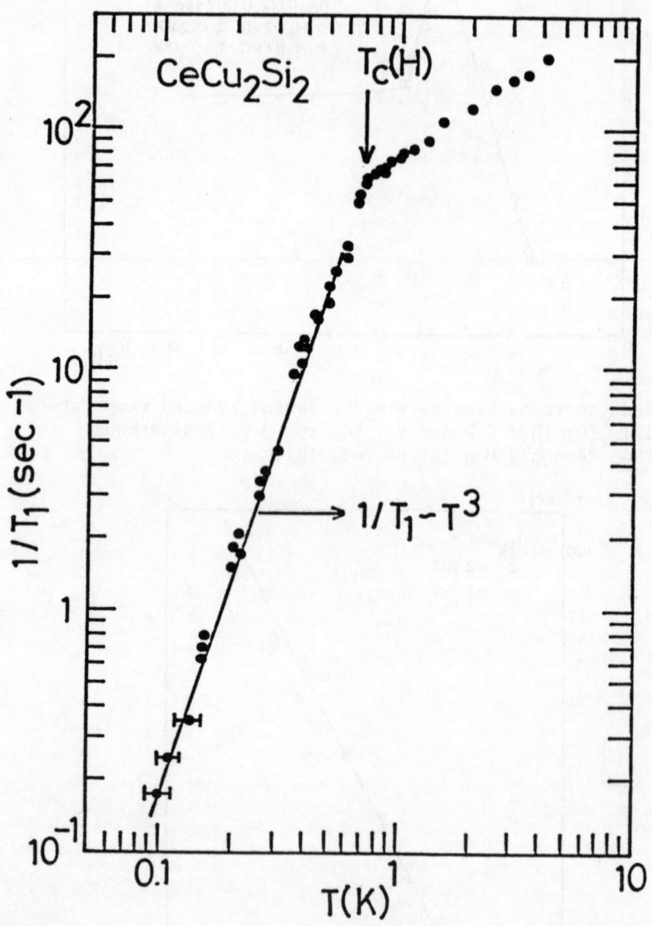

Fig.7. Temperature dependence of the reverse spin-lattice relaxation time of ^{63}Cu in $CeCu_2Si_2$ (H = 5.72 kOe, f = 6 MHz) (after ref. 21).

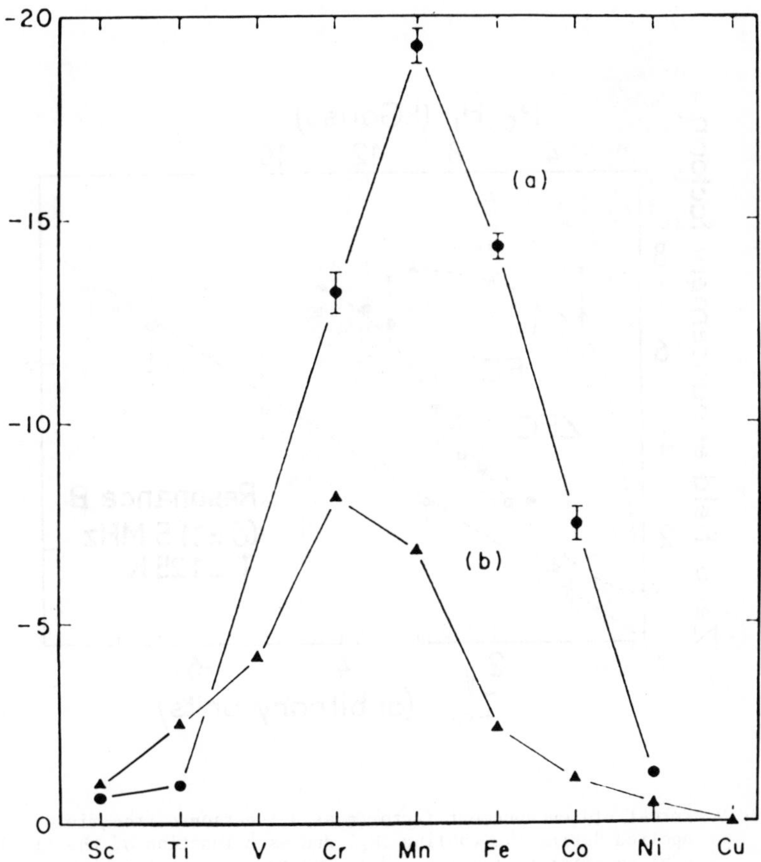

Fig.8. NMR shifts Γ ($\Gamma = \partial \ell nK/\partial x$) for 3d impurities a) liquid Cu and b) liquid Al (after ref. 13).

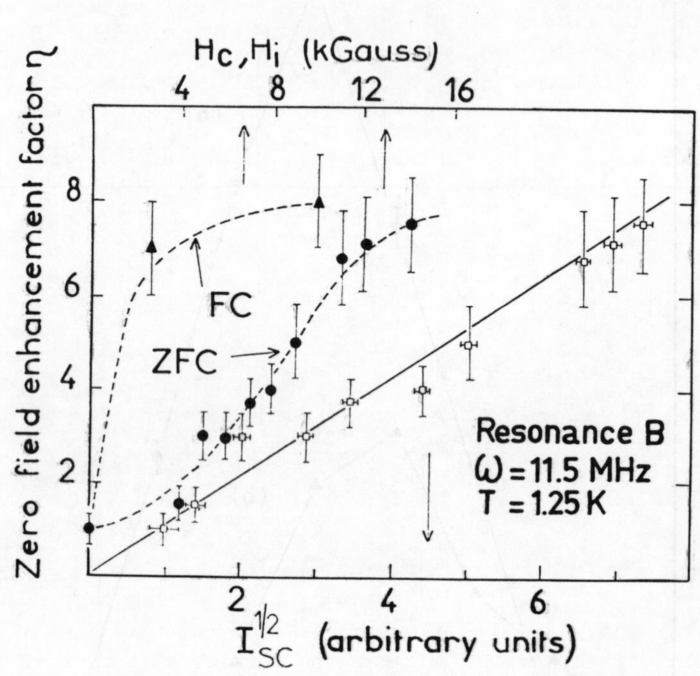

Fig.9. Zero-field enhancement factor η as a function of the field applied during the cooling (H_c) and as a function of the field applied after the cooling (H_i). The factor η scales with $I_{SC}^{1/2}$, I_{SC} being the single-coil signal intensity (after ref. 33).

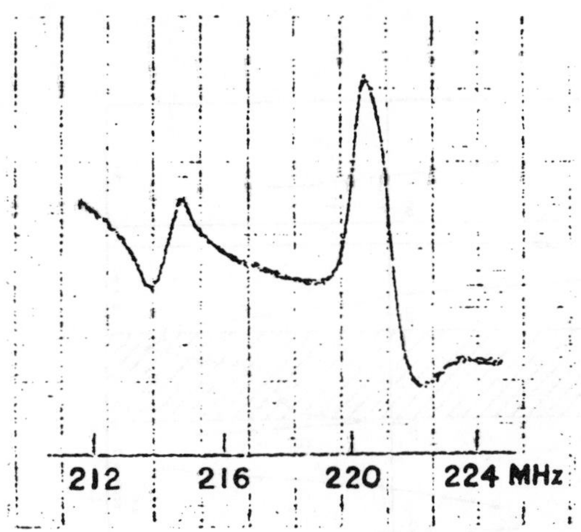

Fig.10. A recorder trace of the zero-field ^{59}Co NMR in hexagonal cobalt at 190 K (detection in quadrature with the modulation) (after ref. 37).

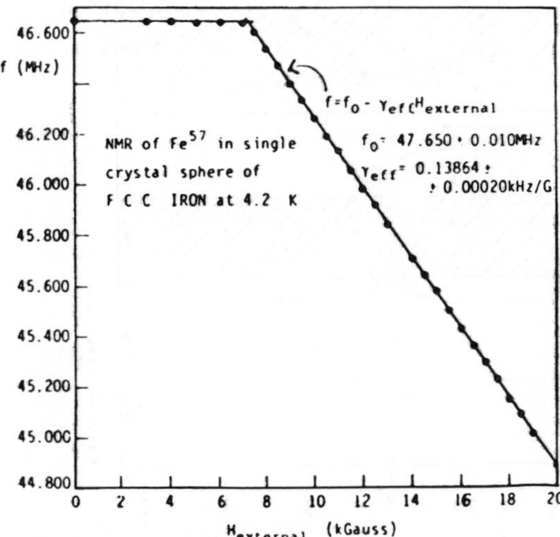

Fig.11. Resonance frequency of ^{57}Fe in a sphere of single crystal bcc iron as a function of external field (after ref. 38).

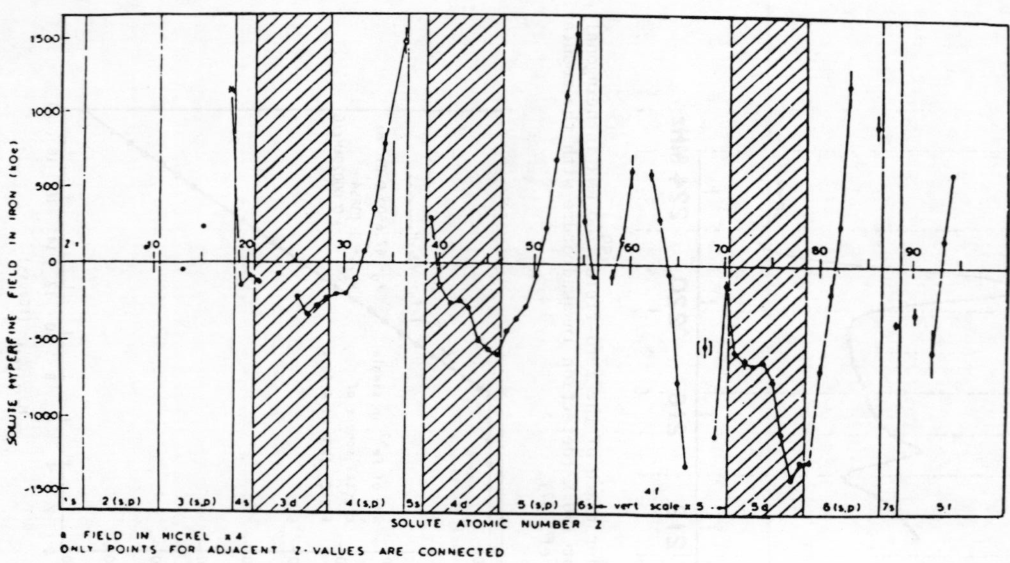

Fig.12. Hyperfine fields for impurities in ferromagnetic iron (after ref. 46).

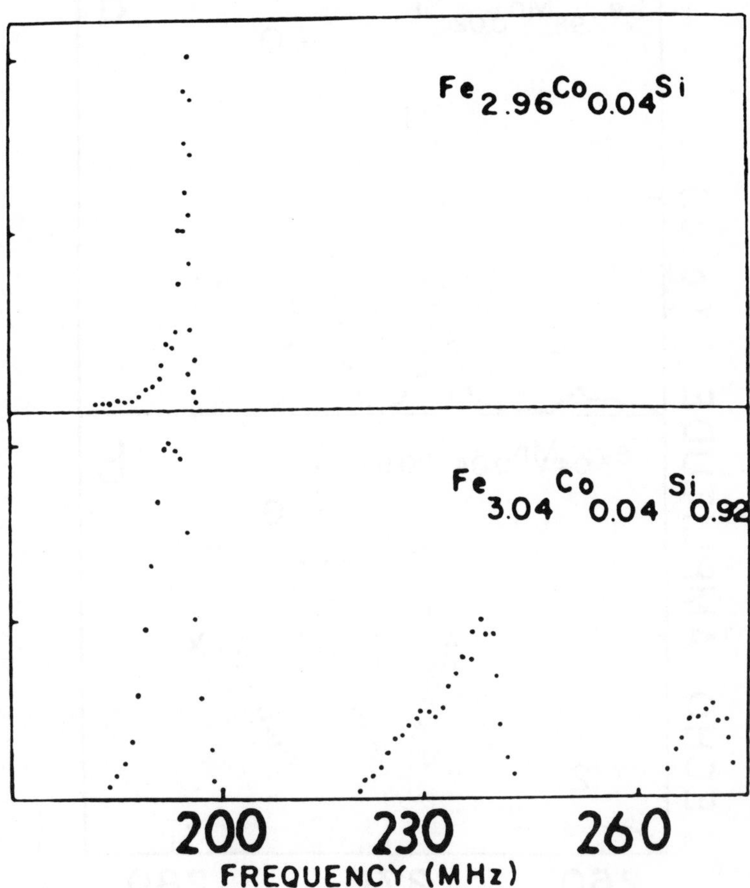

Fig.13. Spin-echo NMR spectra of ^{59}Co in $Fe_{2.90}Co_{0.04}Si$ and $Fe_{3.04}Co_{0.04}Si_{0.92}$. The lines at 192.5, 237.5 and 270.0 MHz are assigned to Co atoms with 4, 5 and 6 Fe 1 nn in the offstoichiometric sample (from ref. 49).

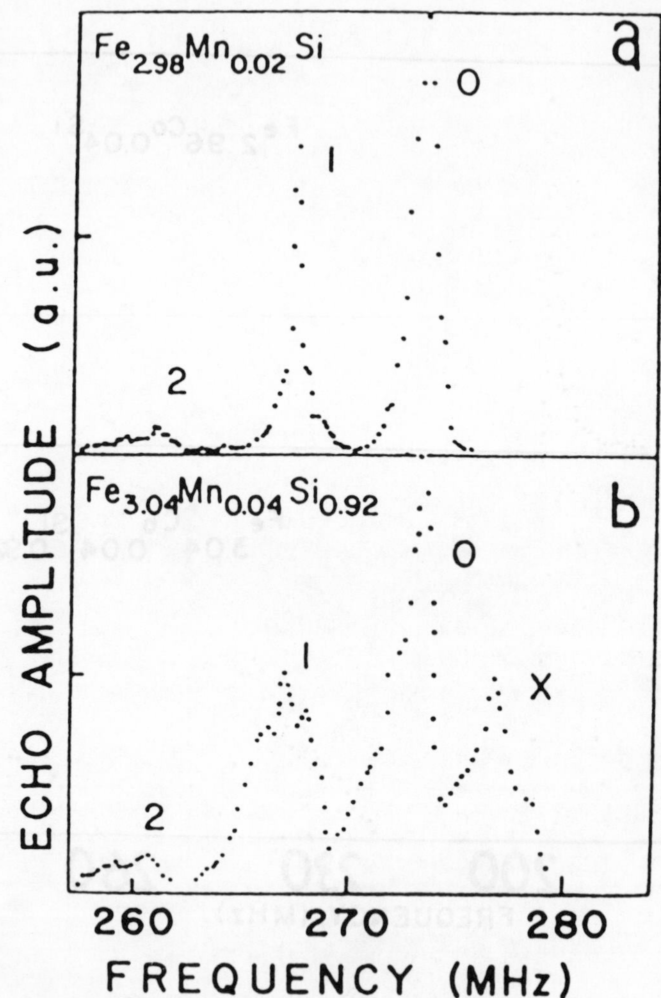

Fig.14. Spin-echo NMR spectra of ^{55}Mn in $Fe_{2.98}Mn_{0.02}Si$ and $Fe_{3.04}Mn_{0.04}Si_{0.92}$. The peaks at 273.2, 267.0 and 260.8 MHz are assigned to Mn with 0, 1 and 2 Mn 2 nn. The peak at 276.8 MHz is from Mn with 1 Fe 2 nn (from ref. 49).

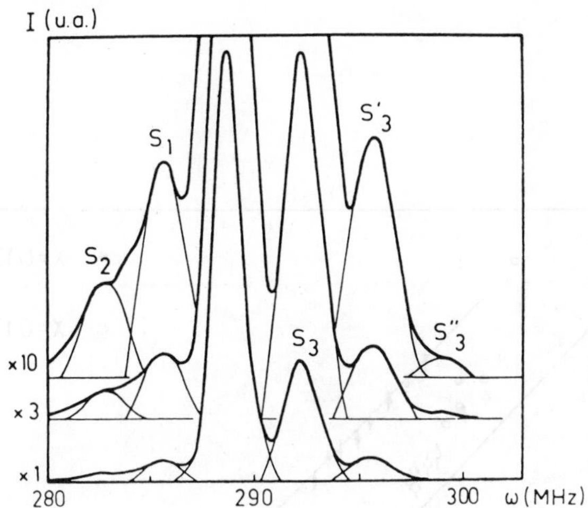

Fig.15. Spin-echo NMR spectra at 1.5 K of ^{59}Co in a Fe-2 at.% Co alloy annealed at 700°C showing discrete satellites on both sides of the main line (288.7 MHz) (from ref. 50).

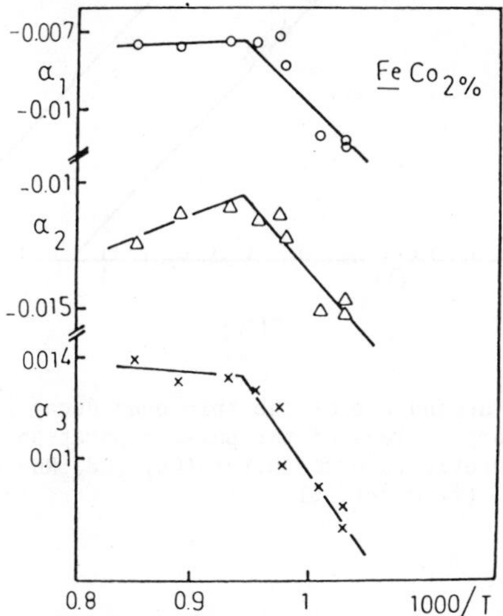

Fig.16. Short-range order parameters α_1, α_2, α_3 as a function of reverse temperature in a Fe-2 at.% Co alloy. Change of slope occurs at the Curie temperature (from ref. 50).

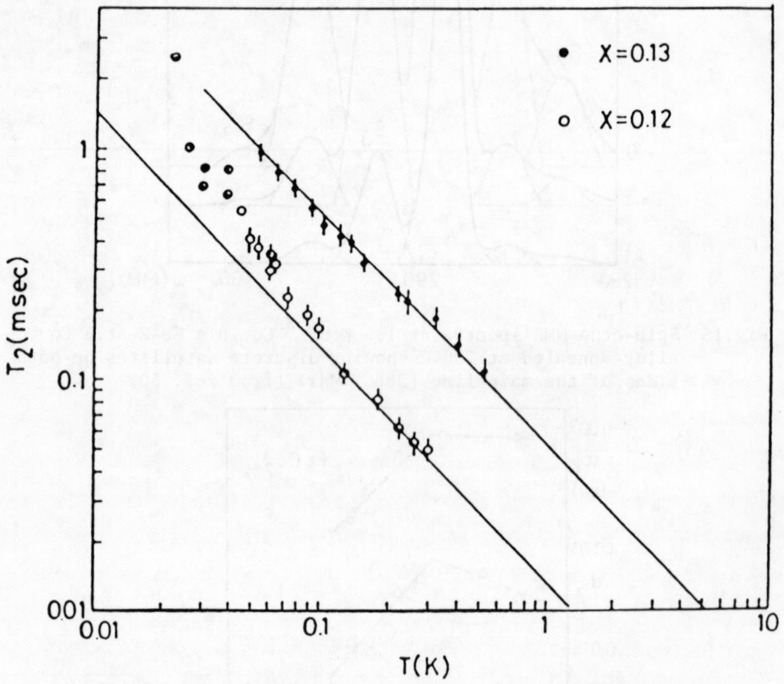

Fig.17. Temperature dependence of the spin echo decay time (T_2) taken for large values of the pulse separation $\tau(T_2(L))$ in the ferromagnetic superconductor $(Ce_{1-x}Gd_x)Ru_2$ with x = 0.12 and x = 0.13 (from ref. 51).

INTRODUCTION TO MAGNETISM OF AMORPHOUS ALLOYS*

A.R. Ferchmin and S. Krompiewski

Institute of Molecular Physics, Polish Academy of Sciences,
Smoluchowskiego 17/19, 60-179 Poznań, Poland

1. INTRODUCTION

This paper does not pretend to over the now vast area of research on physics of magnetically ordered glassy metals and alloys. It represents a rather subjective choice. A remedy for the possible flaws in completeness would perhaps be provided by reading the supplement A to the book "Amorphous Magnetism and Metallic Magnetic Materials - Digest" by A.R. Ferchmin and S. Kobe [1]. This supplement attempts to be complete in subject coverage, and represents an up-to date review of the subject, although most references refer to the papers published in 1981. The most important books on the subject are listed as references [2-10]. Judging from the impact measured by the number of yearly publications, tending to stabilize at the level of some 800 papers a year (cf. ref.[1], Introduction), the amorphous magnetism represents already a nature domain of investigation. The maintaining interest is supported essentially by the starting and prospective applications. Some of the simplest questions remain without answer (e.g.: is the hypothetical amorphous iron magnetic or not?). However, other problems are pursued with real and apparent success, to mention only the progress in understanding

*Part of this paper is based on the lectures given by one of us (A.R.F.) at the Departmento de Fisica of the Universidade Federal de Alagoas, Maceio-Al, Brasil.

the relaxation processes in metallic glasses thanks to Egami and his group [11].

2. STRUCTURE

The usual approach to the atomic structure of solids based on symmetry (translation and point groups) is not applicable to the amorphous. The alternative is to build models. The predictions based on such models can afterwards be compared with experiment. The following should be mentioned:

- Dense Random Packed Hard Sphere (DRPHS) models (mechanical or computer-built). They allow locally fivefold (or other odd-number) symmetry. The right packing density is difficult to attain. Two-component models allow for both atomic and chemical (compositional) disorder. Such models represent a kind of <u>topological</u> disorder, as contrasted to that present in polycrystals or "hot" crystals; they are applicable to metals and ionic compounds.
- Continuous Random Network models. Each site maintains its coordination (defined by chemical bonding), as in the crystalline state, but the structure is built in a random way. "Disclination" lines (going through odd-membered rings of bonds) are present, forming loops or ending at the boundary, as noticed by Rivier [77]. Most applications are to semiconductors, and we leave them out of scope.
- Other models:
  relaxed hard- and soft-sphere computer-built models (von Heimendahl, Boudreaux,...); microcrystalline - like models (Ninomiya, ...; borrowed from dislocational theory of melting, trigonal prismatic packing of Gaskell).

3. STATISTICAL DESCRIPTION OF THE ATOMIC STRUCTURE OF CONDENSED MATTER

Even with a model at hand, one needs parameters, as coordination number, density etc., or building blocks, like elementary cells. Due to the lack of periodicity inherent to the amorphous state, they differ from site to site, and are of statistical character. We can describe the structure using:

- distribution of coordination numbers z (rather than fixed coordination numbers in the crystalline case). They can be found from models, and estimated in some cases from local probes, such as Mössbauer effect or nuclear magnetic resonance, for real materials.
- radial distribution functions, related to two-point distance correlation functions, or, in other words, to the probability of finding an atom at a distance from a given point. Experiments related to the three-point correlation functions for metals are very scarce, and do not concern metallic glasses.
- Bernal (Voronoi) polyhedra - corresponding to the Wigner-Seitz cells in crystals - and their statistics.

4. MAGNETIC MOMENTS OF AMORPHOUS METALS AND ALLOYS

The properties of amorphous cobalt can easily be obtained by extrapolation from its alloys with small quantities of "glass formers", inhibiting the crystallization process. The result is that magnetically the amorphous Co does not practically differ from the crystalline element. This can be supported by the results of the moment method calculations by Khanna et al.[12-13] revealing rather insignificant differences in density of d-states between crystalline (hcp) and model amorphous Co (Fig.1). Similar theoretical evidence is due to Fujiwara[14-15].

Transition metal-metalloid (TM-Me) and "early" transition metal - "late" transition metal alloys (ETM-LTM; i.e., early and late in the periodical table of elements), if their magnetization is plotted versus the total number of valence electrons, reveal the usual Slater-Pauling behaviour (linear in certain ranges of concentration, Fig.2). The earlier interpretations of the Slater-Pauling curve in terms of the rigid--band concept have been abandoned at least from the invention of the coherent potential approximation (CPA), giving evidence on how dramatically the bands can change on alloying. The concept of charge transfer followed. Yamauchi and Mizoguchi [16] supposed for their $TM_{80}B_{10}P_{10}$ (TM = transition metal) alloy a transfer of 1 electron from each B ion and 3 electrons from each P ion to the TM d-band in order to fit the magnetization vs. concentration data. Further on, for alloys containing B or P separately, a transfer of 1.6 electrons from B and 2.4 electrons from P was suggested [17]. Such assumptions enable one to calculate magnetizations for various concentrations with satisfying accuracy; however, it should be stressed, that it is just the linearity of the magnetization vs. valence curves, not the assumption of charge transfer, which makes such calculations effective. Indeed, there exist both experimental and theoretical evidences against transfer in space of such large amount of electrons from metalloid to transition-metal ions.

The experimental evidence of a negligibly small spatial charge transfer comes from photoelectron spectroscopy [18-19] and Compton profile measurements [20]. An explanation not involving essentially the transfer of electrons in space is due to Alben et al. [21]. The s-p electrons of the metalloid ions are hybridized with the electrons of the transition metal elements, with some depolarization of the d-band resulting in reducing the magnetization. Electronic structure calculations by Fujiwara [22] also do not suggest any remarkable charge transfer in space. Small-cluster (Ni_2F_e and Ni_2Fe_2B) calculations by Messmer [23] show that on adding boron the d-band becomes wider and magnetic moment decreases, but there is a very small charge transfer in the direction to boron ions (!).

5. PHENOMENOLOGY OF METALLOID INFLUENCE ON MAGNETISM IN METALLIC GLASSES

Usually, the magnetic moment of a glassy alloy is reduced with respect to its crystalline counterpart, and the reducing is stronger if phosphorus is present than in the case of boron [24]. The empirically found relation that in $Fe_{80}Me_{20-x}Me_x$ a magnetic moment and Curie temperature increase with increasing the atom diameter of the metalloid added, while the inverse is true for $Co_{75}Me_{25-x}Me_x$, is susceptible to a phenomenological interpretation. Fe and Co are supposed to lie on two different sides of the maximum on the Bethe-Slater curve (dependence of the exchange integral J on interatomic distance r), which would give

$$\frac{\partial J(r)}{\partial r} > 0 \text{ for Fe, and } \frac{\partial J(r)}{\partial r} < 0 \text{ for Co.}$$

6. VIRTUAL BOUND (VB) STATES FOR DILUTED ALLOYS AND THE EXTENSION OF FRIEDEL THEORY TO HIGHER CONCENTRATIONS

The strong magnetic behaviour may be characterized by the rule $\frac{\partial m}{\partial n_d} = -1$, i.e., adding of one electron to the d minority band produces a 1 μ_B decrease in magnetic moment μ[11]. Note that the shape of the band can be quite arbitrary, provided that one sub-band is full. Therefore, this rule (as well as the other discussed in this section) apply to both a glassy metal and a crystalline one. If one adds impurities to a metal, resonant or virtual bound states appear in addition to the host band. If the valency (charge) difference $\Delta Z = \delta n_\uparrow + \delta n_\downarrow$ between the host and impurity is larger than two, usually the virtual bound states are pushed out of the host band (Fig.3). Now, for a strong ferromagnet the d↑ electrons lie entirely below Fermi energy, therefore the VB states, if completely pushed out, contribute precisely $\delta n_\uparrow = -5$ per atom to N↑ (the number of spin up electron states). From the fact, that the magnetic moment is a difference N↑ - N↓, there follows the Friedel rule[25]

$$\frac{\Delta\mu}{\mu_B} = \frac{1}{\mu_B}\frac{\partial\mu}{\partial x} = \delta n_\uparrow - \delta n_\downarrow = -(\Delta Z+10)$$

(x - concentration)

valid essentially for dilute alloys. Some amorphous alloys data listed below[24] may exemplify such behaviour:

for V and Nb in Co $\Delta Z = -4$

	$-\mu_B \frac{\partial\mu}{\partial x}$
Friedel theory	6
$Co_{80-x}V_xB_{20}$	4
$Co_{78-x}Nb_xB_{22}$	7

Deviations from theory may be explained by the influence of boron. An excellent review on diluted amorphous alloy moments is due to Durand[26].

It has been recently shown[78,79] that it is practical to extend the traditional meaning of the notion "strong ferromagnetism" (one d sub-band either completely full or empty) to cover the case when a density of states of one of the subbands has a gap (or a pronounced dip) in vicinity of a Fermi energy. Then there is a tendency for the Fermi level to be pinned in the gap and a number of electrons in this sub--band stays almost constant. Likewise in the conventional case, a system behaves as a strong ferromagnet and, in particular, shows a linear change of magnetization upon alloying and its magnetization is hardly sensitive to structural changes and to such external disturbances like pressure and magnetic field.

In a series of papers, Williams, Malozemoff et al.[78,27,80] invented a new elegant and informative way for presentation of magnetic moments of transition metal alloys. The method, referred to as the generalized Slater-Pauling plot construction, is based on the following simple formula

$$\mu = 2n_{\uparrow}^{sp} + Z_m ,$$

where

$$Z_m = 2n_{\uparrow}^{d} - Z$$

is the so-called magnetic valence, Z - the number of valence electrons and $n_{\uparrow}^{sp}$ and $n_{\uparrow}^{d}$ stand for the numbers of up-spin electrons contributed by an sp band and d bands, respectively. The magnetic valence is just the negative of the valence charge except for Fe, Co and Ni for which $2n_{\uparrow}^{d} = 10$ and $Z_m = 2$, 1 and 0. For these transition metals the sp band contributes 0.3 electrons to the up-spin d band, giving $\mu = 2.6$, 1.6 and 0.6 μ_B, respectively. In fact, for Fe $\mu = 2.2\ \mu_B$ since iron is a weak ferromagnet, i.e., the up-spin d band is not completely filled. If a generalized Slater-Pauling plot is a 45° straight line then the alloy may be classified as a strong ferromagnet with constant $n_{\uparrow}^{d}$ and $2n_{\uparrow}^{sp} = \mu(Z_m{=}0)$. A lot of alloys, both crystalline and amorphous, has already been classified in this way (Figs 4-5).

7. SPIN WAVES IN METALLIC GLASSES

The evidence of spin wave excitations in amorphous magnetic systems comes either indirectly, from the temperature dependence of magnetization (Blochs $T^{3/2}$ law), or directly from spin wave resonance, Brillouin scattering [29], and neutron scattering [30-33]. The existence of long periodic spin waves in a non-periodic medium is easy to understand as they do not feel short range inhomogeneities. They follow the usual quadratic dispersion law:

$$E(k) = g\mu_B H + Dk^2 + \text{higher order terms.} \tag{7.1}$$

These excitations contribute to the statistics with the result for magnetization:

$$\frac{M(0) - M(T)}{M(0)} = BT^{3/2} + CT^{5/2} + \delta T^4 + \ldots , \tag{7.2}$$

the coefficient D becoming effectively temperature-dependent:

$$D(T) = D(0)(1-\alpha T^{5/2}). \tag{7.3}$$

The $T^{5/2}$ term in (7.3) has its origin in the magnon-magnon interactions. As regards itinerant electron systems, the theory predicts [81] that a leading temperature-dependent correction to D behaves like T^2 and it is due to the electron-magnon interaction. That is found for crystalline metallic ferromagnets but similar experimental findings have not been yet reported for glassy metals for which the $T^{5/2}$ term seems to be of main importance [82].

A simple formula relates D to the Curie temperature T_c in the Heisenberg model:

$$D = \frac{k_B a^2}{2(S+1)} T_c \tag{7.4}$$

with k_B - Boltzmann constant, a - interatomic distance and S - spin magnitude. An expression following from itinerant-electron ("band") picture also suggests proportionality of D and T_c:

$$D = \frac{\pi}{6\sqrt{2}} \frac{k_B}{k_F^2} T_c \tag{7.5}$$

(k_F - Fermi vector).

Experimental data [34,35] seem to confirm the linear interdependence of D and T_c; however, extrapolations do not usually meet the ($D = 0$, $T_c = 0$) point (Figs 6-7).

It should be added that D values taken from neutron, D_N, (and some other) experiments are usually larger than those derived from the temperature dependence of magnetization (7.2) and Mössbauer experiments, D_M, through the relation of coefficients B and D:

$$B = \zeta(\frac{3}{2}) \frac{g\mu_B}{M(0)} (\frac{k_B}{4\pi D})^{3/2} \tag{7.6}$$

Only a few data could be reconciled by careful experiments on the same samples (examples are given in [36,82]). An extensive table of spin-wave stiffness constant D as measured by various methods can be found in Suppl. A to ref. [1]. In an attempt to explain the discrepancies between D_N and D_M, which sometimes differ by as much as a factor of 2 or 3, one can associate the reduction in magnetization, i.e., a large value of the coefficient B (and small D), with Stoner excitations. An analysis in this spirit has been done for $Fe_xNi_{80-x}B_{18}Si_2$ (x = 15, 20, 40, and 60%) and $Fe_{80}B_{20}$ glassy metals [83]. These systems have been assumed to be strong ferromagnets, a relevant expression for the magnetization is then

$$\frac{M(0) - M(T)}{M(0)} = BT^{3/2} + B_1 T^{3/2} \exp(-\Delta/kT) \; ; \tag{7.7}$$

where Δ is the gap between the top of the full sub-band and the Fermi energy. By fitting this formula to experimental data the parameters B_1 and Δ have been determined, with the result for Δ changing from 2 to 5 meV depending on iron content x. This way to the last term in (7.7) the stiffness constant deduced from magnetization agrees resonably well with that measured directly by neutron scattering.

Spin polarized photoemission measurements on amorphous $Fe_{83}B_{17}$ were made in [84]. The resolution of the experiment did not allow to measure Δ or to reveal negative spin polarization at the Fermi energy typical for strong ferromagnets. However, the obtained spectra were shown to be Ni-like rather than Fe-like and similar to those corresponding to $Co_{78}Mn_2B_{20}$. Thus the results seemed to give some evidence for strong ferromagnetism of the studied alloys. No support was lent to that claim in [85], where $Fe_{82}B_{12}Si_6$, $Fe_{60}Ni_{20}B_{20}$ and $Fe_{44}Ni_{37}B_{19}$ were studied. Those systems were suggested to be weak ferromagnets with a trend towards strong ferromagnetism with increasing Ni content (the spin polarization near the Fermi level got more and more negative). Stoner excitations in glassy metals have already been directly measured by spin polarized electron energy-loss spectroscopy [86]. The observed spin-polarized spectra for $Fe_{82}B_{12}Si_6$ have revealed a peak around the energy loss which corresponds to the exchange splitting of iron. Another way of studying Stoner excitations is by neutron scattering. If a magnon branch merges with a Stoner continuum, intensities of spin wave peaks are expected to decrease rapidly. Such experiments were performed on crystalline Ni and Fe [87,88], where a rapid decrease of spin wave peak intensities at energies close to 100 meV was reported. As it has been shown recently [89] this phenomenon may also be explained for Ni (along the wave-vector direction 100) if one attributes the decrease to the hybridization of the acoustic magnon branch with the optical one. According to this interpretation spin waves persist throughout most of the Brillouin zone, and the Stoner excitations manifest themselves near the zone boundary by causing some extra decrease in intensities. This mechanism offers an explanation of the puzzling observation that the energy at which spin-wave peaks suffer an abrupt drop does not practically depend on temperature, although the theoretical prediction is that it should be proportional to magnetization. So far there have been no neutron scattering measurements of Stoner excitations, analogous to those described above, in amorphous alloys, relevant theoretical considerations on the effect of these excitations on the neutron cross section are also still lacking.

Papers [30-33] revealed an extra branch of short-wavelength spin waves. They were observed around the first peak in the static structure factor $S(q)$, $q = 3.13$ Å^{-1} in Mook's experiments with Co_4P and $q = 3.05$ Å^{-1} for FePC [30,31]. The lowest excitations of this kind should have a large energy gap of some tens of meV according to [30,31]; however, no gap was observed in later experiments [32,33]. A theory based on an amorphous itinerant-electron model has been developed by one of us [37-39] to explain the above mentioned behaviour. The theory is based on a single-band Hubbard hamiltonian and, one-[37,38] or three-dimensional computer-built amrophous structures[39]. The calculation of susceptibility $\chi_{ij}(E)$ in the ladder approximation has led to a spectral density function clearly showing a gapless short-wave-length spin wave branch in the vicinity of the first peak in the static structure factor, in agreement with the recent experiments[32,33]. Yet another approach to the problem of spin waves in amorphous magnets is worth mentioning. Ninomiya observed, that some Bernal cells correspond to atomic configurations in the cores of dislocations in some crystal structures[40,41]. For example, Bernal's tetragonal dodecahedra arise in the cores of screw dislocations in fcc structures, or in edge (60°) dislocations. Another relevant fundamental units is a trigonal prism, capped with half octahedra or not. A dislocation model of amorphous metals arose based on this observation[41]. A Heisenberg model study of the spin-wave density of states (DOS) changes due to the topological disorder alone (no change in the strength of interaction was assumed) is due to Schmidt[42]. On applying the moment method (8-1o moments) he was able to demonstrate that most changes in DOS occur at the top of the spin-wave band.

8. RANDOM ANISOTROPY

Explaining the properties of rare earths (RE) and their alloys requires in general taking into account their inherent large local anisotropy. The local anisotropy is mostly of one-ion type and comes from the electrical field ("crystal field") of the neighbouring ions acting on rare-earth electron orbitals, influencing their spin-dependent part via the spin-orbit coupling. It can be described by an effective Hamiltonian H, containing in addition to the usual exchange part

$$H_e = - I \sum_{ij} \vec{J}_i \cdot \vec{J}_j \tag{8.1}$$

($\vec{J}$ - total moment of the ion) an extra anisotropy term

$$H_c = \sum_i D_i (\vec{J}_i \cdot \vec{n}_i)^2 , \tag{8.2}$$

$$H = H_e + H_c , \tag{8.3}$$

where I denotes an exchange parameter, and $\vec{D}_i$ and $\vec{n}_i$ represent the magnitude and direction, respectively, of the local one-ion magnetic anisotropy[1]. In amorphous substances, the direction of the local anisotropy varies from site to site in a random way, and its value can even be enhanced with respect to that in a crystal due to the lack of symmetry producing usually some cancellation among various contributions. Then, (8.3) becomes the Harris-Plischke-Zuckermann (HPZ) [45)] hamiltonian. The consequences of randomization of the local uniaxial anisotropy are multiple, from a variety of new magnetic phases, to reduction in total magnetization and transition temperatures. Cases where no fit to the HPZ model with ferromagnetic interactions could be achieved (RE-Cu [46)]), have been explained in a two-point approximation with antiferromagnetic coupling [47)].

8.1 Types of Magnetic Ordering in RE-Containing Amorphous Alloys

Various types of magnetic ordering are shown schematically in Fig.8, where magnetic moments are plotted from one common origin instead of from their actual positions. They are exemplified by ferromagnetic YCo a, speromagnetic TbAg b, asperomagnetic YFe and DyNi (c and d, respectively; we shall explain further on the different origin of TM moments "fanning" in the former and RE "fanning" in the latter). The last four examples represent sperimagnetic ("amorphous ferrimagnetic") types of ordering for alloys with heavy RE (DyCo e and DyFe f) and light RE (NdCo g and NdFe h). The terms like "speromagnetic" have been proposed by Coey [48)].

8.2 Magnetic Properties of Random Anisotropy Magnets

The original HPZ theory explained qualitatively - in molecular field approximation - the reduction of T_c and saturation magnetization in amorphous $TbFe_2$ (with respect to the crystalline phase) as a result of competition between exchange and anisotropy energies, as characteriz-

[1]Note that we have used here for simplicity the anisotropy of the $(J^z)^2$ type. This restriction can be lifted[43,44)]. Do not confuse D_i with the spin-wave stiffness constant D!

ed by the ratio $\varepsilon = D/z_{RE}I$ (z_{RE} - number of RE nearest neighbours). For this model, the classical energy of the magnetic moment J_i is [49]:

$$E_i = -gI\mu_B(H+H_W)\cos\psi_i - D(J^{z_i'})^2\cos^2(\theta_i-\psi_i) \qquad (8.4)$$

where H - external field, H_W - Weiss molecular field (both of direction z), θ_i - angle between the z-axis and local anisotropy direction z_i', ψ_i - angle between the z-axis and local moment direction z_i (Fig.9). From (8.4), for dominant exchange $\varepsilon \simeq 0$, there follows a complete alignment of moments ($m \simeq 1$, $\psi \simeq 0$), or an asperomagnetic state. For diminant anisotropy $\varepsilon \to \infty$, $m \simeq 0.5$, $-\frac{\pi}{2} < 4 < \frac{\pi}{2}$. Note that since (8.4) is classical, it applies to ions of large moment J, such as Dy ($J = \frac{15}{2}$). Let us analyze the situation in some details:

i) Dominant exchange

The alloy can order ferromagnetically. We encounter the situation common to metallic glasses, as discussed in previous chapters. Use of itinerant-electron approach is suggested rather than Handrich's formula [2], exploiting the concept of a spread ΔI in the exchange integral values to explain flattening of the M(T)/M(0) vs. T/T_c curve (Fig.10). Some criticism concerning the Handrich's formula may be found in [1,11] and [50]. Another possible kind of ordering is the asperomagnetic one. The susceptibility curve of GdAg [51] with T_c = 122 K and the paramagnetic Curie temperature θ_c = 130 K, but saturation at fields as high as ∿100 kOe, suggests that it falls into this category. Taking into account the S-state ion character of Gd, the random anisotropy cannot be responsible for this ordering and rather a small portion of antiferromagnetic exchange in the ΔI distribution should be invoked [44]. While it is probably characteristic also of YFe for some range of concentration to be asperomagnetic (for lower Fe concentration it becomes speromagnetically ordered [49], (Fig.11). Sellmyer and O'Shea [44] classify the behaviour of a similar exchange-fluctuation-dominated alloy containing Gd as a single magnetic ion ($Gd_xLa_{72-x}Ga_{18}B_{10}$) as speromagnetic (<u>not</u> asperomagnetic), and propose a phase diagram with a spin glass phase resembling that stemming from the Sherrington-

-Kirkpatrick theory [52,53].

ii) Dominant anisotropy

Let us consider the heavy RE first. TbAg represents an example; it does not saturate even at 400 kOe [54] (Fig.12). The single-ion anisotropy dominates with random spatial distribution of the local easy axes, resulting in an asperomagnetic order with mainly ferromagnetic exchange and a small amount of the antiferromagnetic exchange. Further example: $Dy_{21}Ni_{79}$ (see, e.g., [1] and references therein). A local probe (Mössbauer effect) gives the moment of 10.7 μ_B per Dy ion, close to 10.1 μ_B for the free Dy ion (in Ag), while from magnetization the average moment $\bar{\mu} = 5.1\ \mu_B$. For $\psi = \frac{\pi}{2}$ (Dy moment directions scattered within a hemisphere), Chappert [49] proposes $\bar{\mu} \quad \mu_{Dy}\ \overline{\cos\psi} = \mu_{Dy} \cdot \frac{1}{2}(1+\cos\psi)$, giving roughly the latter value for $\bar{\mu}$. The light RE (Nd, Pr) alloys may be characterized by mainly antiferromagnetic coupling with a few ferromagnetic interactions, but essentially the same resulting magnetic structures (Fig.8).

iii) Exchange comparable to anisotropy

For $Co_{77}Dy_{23}$ no saturation of magnetization up to 150 kOe is attained; a compensation temperature of 230 K can be observed on the M(T) curve. Mössbauer experiments suggest sperimagnetic ordering with the cone half-angle of $\psi \simeq 70^{o}$ [55]. This can be explained with strong Co-Co exchange, small negative CoDy exchange, putting the Dy moments to the inverse direction, and large Dy-Dy anisotropic interactions that scatter its directions in space.

iv) Spin waves in RE-containing amorphous alloys

Chappert [49] stresses that very few observations have been reported on spin waves in the RE-based alloys, quoting the $T^{3/2}$ law found for magnetization in $GdAl_2$ [56]. However, a number of spin wave resonance experiments has been performed on GdCoMo and GdCo amorphous films [57-60,90-92].

9. DILUTION IN AMORPHOUS FERROMAGNETS

Most of theories are confined to random dilution of magnetic systems with regular lattices, with a random interaction added [61] or not, or an irregular network with the coordination number z identical for all atoms [62], which amounts to more or less the same thing. Only recently there appeared a couple of papers dealing with locally fluctuating coordination z [63] or dilution in a model amorphous structure [64,50,3]. Yet another approach is based on a model lattice that is a self--similar fractal gasket. The problem should be split into two parts. Either it concerns localized-electron systems (insulators, f-electron systems ...), where percolation behaviour dominates, or itinerant-electron systems whose behaviour is dominated by other processes (to quote only the amorphous FeGe with its change from magnetic metal to nonmagnetic nonmetal at a certain concentration). Here, we confine ourselves to the former.

Theories based on the concept of disorder with regular underlying lattices may reproduce the overall shape of the curve T_c versus concentration more or less correctly. However, they cannot guarantee the reproduction of the correct position of the intersection of this curve with the concentration axis, i.e. the critical concentration. The transition in this point is of the nature of a percolation threshold. It is well known that percolation depends only on the topological properties of the network, i.e. its connectivity, but <u>not</u> on the strengths of the bonds. It follows that a theory invoking a lattice or Continuous Random Network model, being unable to reproduce the kind of connectivity inherent in a DRPHS model (now believed to resemble that existing in metallic glasses), cannot pretend to give the correct critical concentration of amorphous magnetics.

The authors of [64,50] took a DRPHS model of about 1000 spheres computer-built by Finney, chose some of them in a random way as nonmagnetic, and calculated the probabilities and mean concentration numerically. These data were then used to calculate the Curie temperature of the system, within a scheme due to Matsudaira [66], as generalized for arbitrary coordination in Ref. [67]. The reader is referred to the original publication for the details of this method; here we simply stress that, unlike many effective-field theories, it leads to equations which are algebraic and not transcendental in the spin variables, and that it proved to give reasonable values of the critical concentrations for various systems [68]. The main result is a curve of T_c' versus concentration which is very typical for percolation processes (Fig.13). The plot shows a linear part and a part with a steep decrease to zero at the critical concentration value c_o. For an interaction range of 1.5 (with the sphere diameter being taken as unity) the critical concentration c_o amounts to about 0.15; for comparison in the same approximation one gets 0.13 for an fcc lattice. This should be compared with the value of the percolation concentration $c_o = 0.17$ obtained by Powell [69] for another DRPHS system and the same interaction range of 1.5. For comparison of Fig. 13 with results of other theories see [3].

Experimental studies on magnetic dilution in amorphous materials concern mainly metallic alloys. The earliest of these materials was Fe-Pd-P [70] with a snowslide form of the T_c versus c curve, possibly due to additional second-neighbour interaction (compare with Ref. [71]). Yet another experimental plot (Fig.14, cf. also Refs [72,73,3] agrees only qualitatively with that predicted in [64,50].

Now, in various experiments c_o ranges from some 12% to 60%. It is a real challenge to try to reconcile the experimental and theoretical c_o values, even if one is persuaded that only a part of them is due to percolation. One can play with parameters. We have already mentioned the interaction range. Another possibility represents the local order and fluctuating coordination number z. With the help of three parameters: mean square fluctuation of z, $\delta = (\langle z_A \rangle - \langle z_B \rangle)/z_A$, describing the chemical short range order (SRO), and α , representing the generalization of Cowley's atomic SRO parameter for the amorphous [74], Maksymowicz [63] was able to predict the critical concentrations for magnetism c_o from zero to 0.595. It has been done on applying the Bethe--Peierls-Weiss and Green functions (with Tyablikov or Hartree-Fock-like decoupling) methods to the Ising model of an amorphous diluted magnet with SRO. However, he did not compare in detail his results with those experiments, where SRO in amorphous metallic glasses was studied.

Up to now, we have discussed the appearance of magnetic order in the diluted systems. Now, once the magnetic order sets in, what about excitations near the percolation threshold? This question represents the subject of a recent paper by Meichle and Salamon [75] on $(Fe_pNi_{1-p})_{80}P_{20}$. From temperature dependence of magnetization in this metallic glass they derived, basing on eq. (7.2), the slope of B raising to infinity, and T_c, saturation magnetization M(0), and spin-wave stiffness constant D, all tending to zero with iron concentration p tending to the critical value of $p_o = 0.14$. They related their results to the scaling arguments based on a self-similar fractal gasket presented in [76], and arrived at a scaling relation.

$$\frac{D(p)}{T_c(p)} = (p-p_o)^{-0.50}$$

which favourably compared with their experiments. Note, however, that this comparison required at least an fcc-lattice connectivity of the amorphous alloy ($p_o = 0.14$ is the value for fcc lattice with nnn interactions).

ACKNOWLEDGEMENTS

One of us (ARF) would like to thank the Universidade Federal de Alagoas for the opportunity of giving the lectures forming a base of this text and the CNPq for a travel grant. He is particularly indebted to Professor E.F. Sarmento for his invitation to come to Macei'o and his warm hospitality. His thanks go also to the staff of the Departamento de Fisica of this University for the time they devoted to him and their help in the endless problems arising.
This work was supported by the Polish Academy of Sciences under Project No. MR-I.9.

REFERENCES

1) Ferchmin, A.R., Kobe, S., Amorphous Magnetism and Metallic Magnetic Materials - Digest (North-Holland, Amsterdam, 1983); Supplement A, Kobe, A., Ferchmin, A.R., Nose, H., Stobiecki, F., J. Magn.Magn. Mater., to be publ.

2) Handrich, K., Kobe, S., Amorphe Ferro - und Ferrimagnetika (Akademie Verlag, Berlin (Physik Verlag, Weinheim, 1980). In German.
3) Kaneyoshi, T., Amorphous Magnetism (CRC Press, Boca Raton, 1984).
4) Hasegawa, R. (ed.), Glassy Metals: Magnetic, Chemical and Structural Properties (CRC Press, Boca Raton, Florida, 1983).
5) Luborsky, F.E. (ed.), Amorphous Metallic Alloys (Butterworths,1983).
6) Koorjani, K., Coey, J.M.D., Magnetic Glasses (Elsevier, Amsterdam, 1983).
7) Güntherodt, H.-J., Beck, H. (eds.), Glassy Metals I and II (Springer Verlag, Berlin, 1981, 1983).
8) Chikazumi, S., Physics of Ferromagnetism. New Edition. Vol. I 1978, Vol. II 1984 (Shokabo, Tokyo, 1978 and 1984). In Japanese, Russian translation of Vol.I now available.
9) Burzo, E., Fizica Fenomenelor Magnetic, Vols. I-III (Editura Academiei, Bucharest, 1981-1983). In Rumanian.
10) Anantharaman, T.R. (ed.), Metallic Glasses: Production, Properties and Applications (Trans Techpublications, Aedermannsdord, 1984).
11) Egami, T., Rep. Prog. Phys., 47, 1601 (1984).
12) Khanna, S.N., Cyrot-Lackmann, F., Dejongueres, M.C., J. Phys. F 9, 79 (1979).
13) Khanna, S.N., Cyrot-Lackmann, F., Phys. Rev. B 21, 1412 (1980).
14) Fujiwara, T., J. Phys. F 12, 661 (1982).
15) Fujiwara, T., J. Non-Cryst. Solids 61/62, 1039 (1984).
16) Yamauchi, K., Mizoguchi, T., J. Phys. Soc. Jpn. 39, 541 (1975).
17) O'Handley, R.C., Hasegawa, R., Ray, R., Chou, C.-P., Appl. Phys. Lett. 29, 330 (1976).
18) Amamou, A., Krill, G., Solid State Commun. 31, 971 (1979).
19) Amamou, A., Krill, G., Solid State Commun. 33, 1087 (1980).
20) Suzuki, K., Itoh, F., Misawa, M., Matsuura, M., Fukunaga, T., J. de Physique Colloque 41, C8-179 (1980).
21) Alben, R.A., Budnick, J.I., Cargill III, G.S., in: Metallic Glasses, eds J.J. Gilman and H.J. Leamy (ASM, Metals Park, Oh., 1978), p.304.
22) Fujiwara, T., J. Phys. F 12, 661 (1982).
23) Messmer, R., Phys. Rev. B 23, 1616 (1981).

24) O'Handley, R.C., in ref. 5.

25) Friedel, J., Nuovo Cim. 7, Suppl. 2, 287 (1958).

26) Durand, J. in ref. 4.

27) Williams, A.R., Moruzzi, L.V., Malozemoff, A.P., Terakura, K., IEEE Trans. Magn. MAG-19, 1983 (1983).

28) Terakura, K., Kanamori, J., Progr. Theor. Phys. 46, 1007 (1971).

29) Malozemoff, A.P., in ref. 7.

30) Mook, H.A., Wakabayashi, N., Pan, D., Phys. Rev. Lett. 34, 1029 (1975).

31) Mook, H.A., Tsuei, C.C., Phys. Rev. B 16, 2184 (1977).

32) Shirane, G., Axe, J.D., Majkrzak, C.F., Mizoguchi, T., Phys. Rev. B 26, 2575 (1982).

33) McK.Paul, D., Cowley, R.A., Stirling, W.G., Cowlam, N., Davies, H. A., J. Phys. F 12, 2687 (1982).

34) Luborsky, F.E., Walter, J.L., Liebermann, H.H., Wohlfarth, E.P., J. Magn. Magn. Mater. 15-18, 1351 (1980).

35) Kaul, S.N., Phys. Rev. B 27, 5761 (1983).

36) Kaul, S.N., Phys. Rev. B 24, 6550 (1981).

37) Krompiewski, S., Phys. Rev. Letters 51, 1092 (1983).

38) Krompiewski, S., Acta Magnetica (Poznan, Poland) I, 3 (1984).

39) Krompiewski, S., Acta Magnetica (Poznan) II,69 (1985)

40) Ninomiya, T., Proc. Symposium on the Structure of Non-Crystalline Materials (Taylor and Francis, London, 1977) p.45.

41) Koizumi, H., Ninomiya, T., J. Phys. Soc. Jpn. 49, 1022 (1980).

42) Schmidt, W., Acta Phys. Pol. A68, 231 (1985).

43) Mukamel, D., Grinstein, G., Phys. Rev. B 25, 381 (1982).

44) Sellmyer, D.J., O'Shea, M.J., J. Less-Common Metals 94, 59 (1983).

45) Harris, R., Plischke, M., Zuckermann, M.J., Phys. Rev. Lett. 31, 160 (1973)

46) Heiman, N., Kazama, N., J. Appl. Phys. 49, 1686 (1978).

47) Bhattacharjee, A.K., Coqblin, B., Solid State Commun. 17, 599 (1978).

48) Coey, J.M.D., J. Appl. Phys. 49, 1646 (1978).

49) Chappert, J., in: Magnetism of Metals and Alloys, ed. M. Cyrot (North-Holland, Amsterdam, 1982) p. 487.

50) Ferchmin, A.R., in: Magnetism in Solids. Some current topics. Proceedings of the Twenty Second Scottish Universities Summer School in Physics, Dundee, August 1981. A. NATO Advanced Study Institute. Eds. A.P. Cracknell and R.A. Vaughan (Scottish Universities Summer School in Physics, Edinburgh, 1981) p. 61.

51) Boucher, B., IEEE Trans. Magn. MAG-13, 1601 (1977).

52) Sherrington, D., Kirkpatrick, S., Phys. Rev. Lett. 35, 1792 (1975).

53) Sherrington, D., Kirkpatrick, S., Phys. Rev. B 17, 4394 (1978).

54) Boucher, B., IEEE Trans. Magn. MAG-13, 1601 (1977).

55) Coey, J.M.D., Chappert, J., Rebouillat, J.P., Wang, T.S., Phys. Rev. Lett. 36, 1061 (1976).

56) Coey, J.M.D., von Molnar, S., Gambino, R.J., Solid State Commun. 24, 167 (1977).

57) Mizoguchi, T., Kobliska, R.J., Maekawa, S., Imazu, T., J. Appl. Phys. 50, 1626 (1979).

58) Prasad, S., Krishnan, R., Suran, G., Sztern, J., Jouve, J., Meyer, R., J. Appl. Phys. 50, 1683 (1979).

59) Rusov, G.I., Zherikhov, S.P., Bochkarev, V.F., Torba, F.G., Fizika Magnitnykh Plenok (Irkutsk), 1979, 13, p. 37. In Russian.

60) Torba, G.F., Zherikhov, S.P., Rusov, G.I., Serebryakov, S.A., Fiz. Metal. Metalloved 51, 669 (1981). In Russian.

61) Medvedev, M.V., Fiz. Tverd. Tela 22, 1944 (1980). In Russian.

62) Moorjani, K., Ghatak, S., J. Phys. C 10, 1027 (1977) and references therein.

63) Maksymowicz, A.Z., Phys. Stat. Sol. (b) 122, 519)1984).

64) Nazarewicz, W., Zagorski, A., Ferchmin, A.R., Kobe, S., Phys. Stat. Sol. (b) 106, K131 (1981).

65) Harris, C.K., Stinchcombe, R.B., Phys. Rev. Lett. 50, 1399 (1983).

66) Matsudaira, N., J. Phys. Soc. Jpn. 35, 1593 (1973).

67) Kobe, S., Ferchmin, A.R., Szlaferek, A., Acta Phys. Polon. A 55, 707 (1979).

68) Ferchmin, A.R., Maciejewski, W., J. Phys. C 12, 4311 (1979).

69) Powell, M.J., Phys. Rev., B 21, 3725 (1980).

70) Sharon, T.E., Tsuei, C.C., Phys. Rev. B 5, 1047 (1972).

71) Matsudaira, N., Takase, S., J. Phys. Soc. Jpn. 36, 305 (1974).

72) Kazama, N.S., Heiman, N., Watanabe, H., Sci. Rep. Res. Inst. Tohoku Univ., Suppl. March p.131 (1978).

73) Heiman, N., Kazama, N., J. Appl. Phys. 49, 1816 (1978).

74) Maksymowicz, A.Z., Phys. Stat. Sol. (b) 114, 125 (1982).

75) Meichle, L.S., Salamon, M.B., J. Appl. Phys. 55, 1817 (1984).

76) Harris, C.K., Stinchcombe, R.B., Phys. Rev. Lett. 50, 1399 (1983).

77) Rivier, N., Simposio Latinoamericano de Fisica dos Sistemas Amorfos, Niteroi, 1984, preprint.

78) Malozemoff, A.P., Williams, A.R., Moruzzi, V.L., Phys. Rev. B 29, 1620 (1984).

79) Malozemoff, A.P., Williams, A.R., Moruzzi, V.L., Terakura, K., Phys. Rev. B 30, 6565 (1984).

80) Williams, A.R., Malozemoff, A.P., Moruzzi, V.L., Matsui, M., J. Appl. Phys. 55, 2353 (1984).

81) Marshall, W., Lovesey, S.W., Theory of Thermal Neutron Scattering (Clarendon Press, Oxford 1971).

82) Birgenau, R.J., Tarvin, J.A., Shirane, G., Gyorgy, E.M., Sherwood, R.C., Chen, H.S., Phys. Rev. B 18, 2192 (1978).

83) Babic, E., Marohnic, Z., Wohlfarth, E.P., Phys. Lett. 95 A, 335 (1983).

84) Allenspach, R., Colla, E., Mauri, D., Landolt, M., Phys. Lett. 105 A, 145)1984).

85) Hopster, H., Kurzawa, R., Raue, R., Schmitt, W., Güntherodt, G., Walker, K.H., Güntherodt, H.-J., J. Phys. F 15, L11 (1985).

86) Hopster, H., Raue, R., Clauberg, R., Phys. Rev. Lett. 53, 695 (1984).

87) Mook, H.A., Nicklow, R.M., Thompson, E.D., Wilkinson, M.K., J. Appl. Phys. 40, 1450 (1969).

88) Mook, H.A., Nicklow, R.M., Phys. Rev. B 7, 336 (1973).

89) Cooke, J.F., Blackmann, J.A., Morgan, T., Phys. Rev. Lett. 54, 718 (1985).

90) Maksymowicz, L.J., Sendorek, D., J. Magn. Magn. Mater. 37, 177 (1983).

91) Maksymowicz, L.J., Sendorek, D., Zuberek, R., J. Magn. Magn. Mater. 46, 295 (1985).

92) Maksymowicz, L.J., Sendorek, D., Zuberek, R., Thin Solid Films 127, 123 (1985).

93) Hauser, J.J., Phys. Rev. B 12, 5160 (1975).

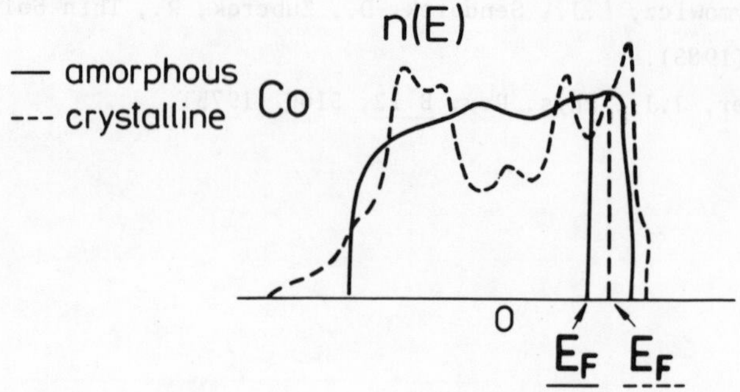

Fig.1. Density of electronic d-states for crystalline and amorphous cobalt (after Ref. [12-13]).

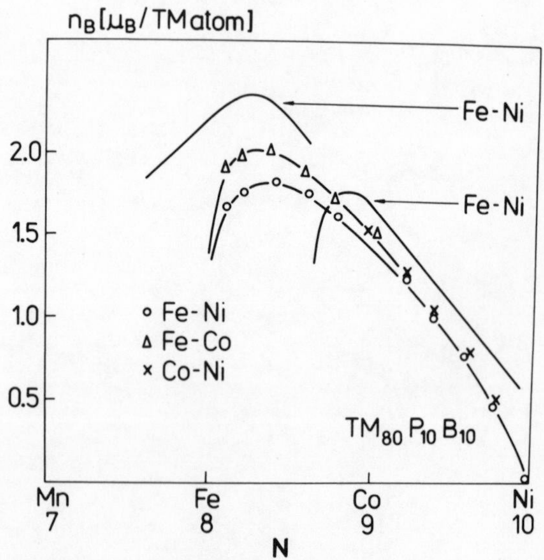

Fig.2. Slater-Pauling plot for some amorphous alloys as compared with crystalline Fe-Ni Invar alloy.

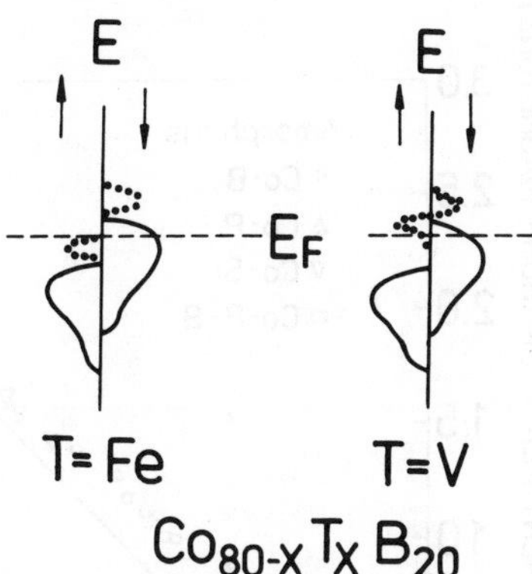

Fig.3. Virtual bound states in T-metal substituted $Co_{80}B_{20}$ alloy.

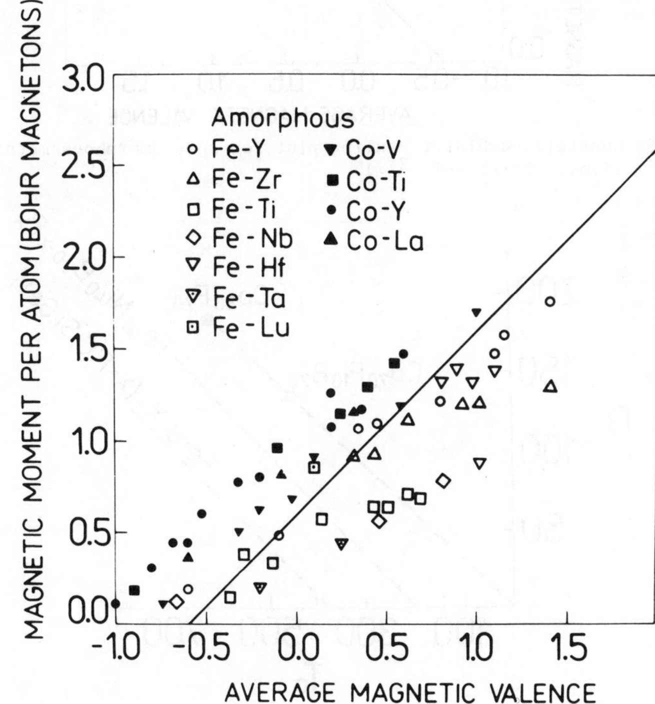

Fig.4. Generalized Slater-Pauling plot for some amorphous metal-metalloid alloys (after Ref. [27]).

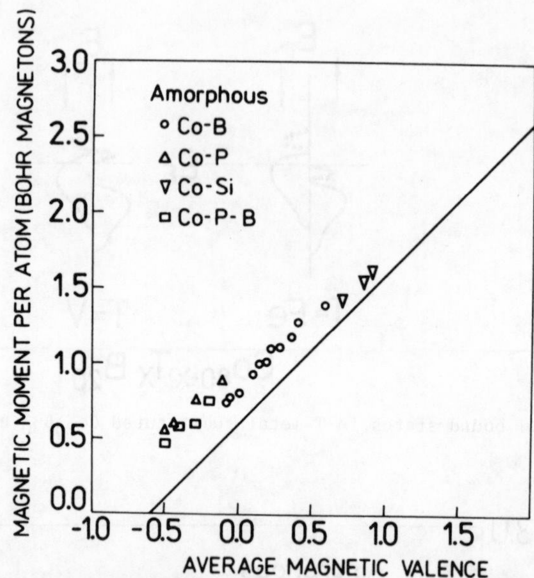

Fig.5. Generalized Slater-Pauling plot for some amorphous metal-metal alloys (after Ref. [27]).

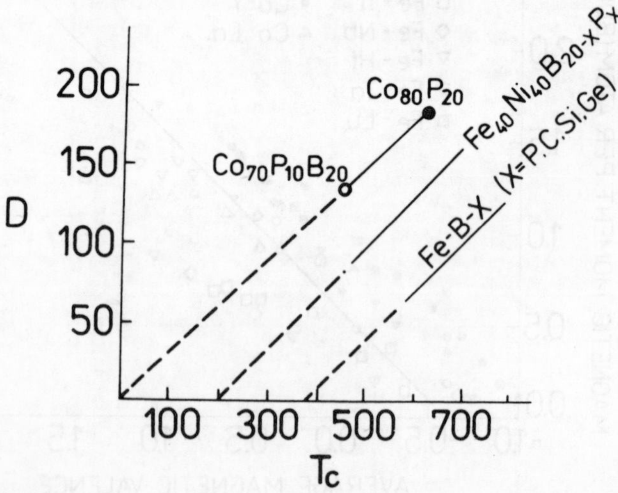

Fig.6. Spin-wave stiffness constant D as a function of Curie temperature T_C for some amorphous alloys (cf. Ref. [34]).

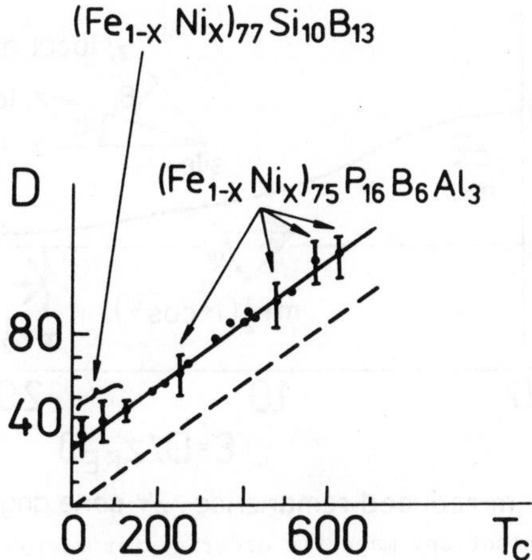

Fig.7. Spin-wave stiffness constant D versus Curie temperature T_c for amorphous Fe-Ni-Me alloys (after Ref. [35]).

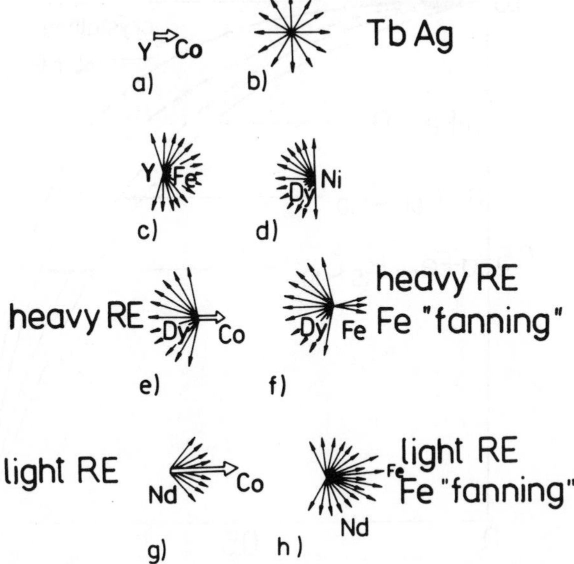

Fig.8. Various types of magnetic ordering. The arrows symbolizing local magnetic moments are shifted to one common origin. a) ferromagnetic, b) speromagnetic, c) and d) asperomagnetic, e) - h) sperimagnetic.

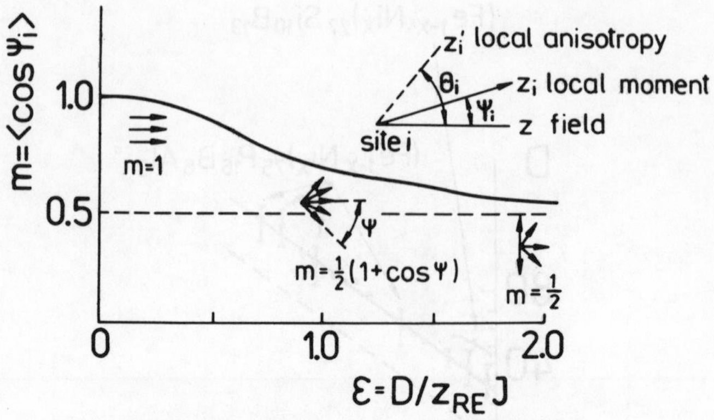

Fig.9. Random anisotropy magnetic ordering for various anisotropy to exchange ratios in a classical approximation (after Ref.[49]).

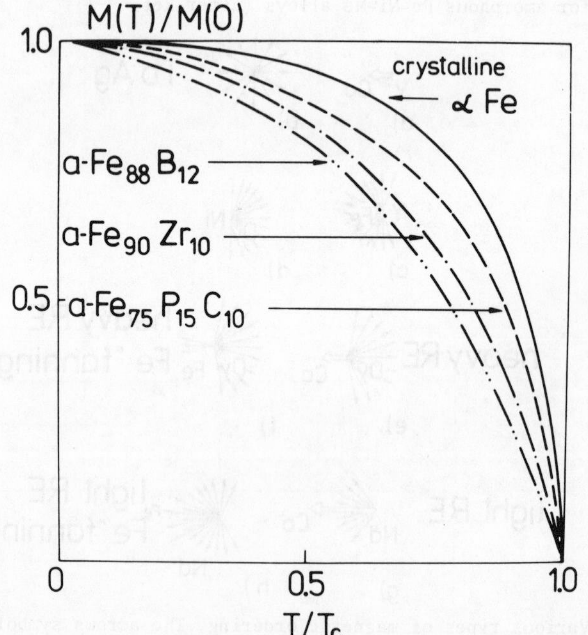

Fig.10. Magnetization vs. temperature plots for some amorphous alloys showing a flattening with respect to the curve for non-alloyed α-iron.

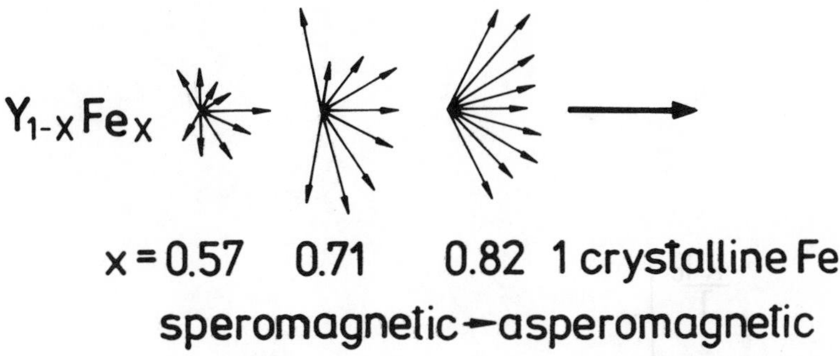

Fig.11. Types of magnetic ordering of amorphous Y-Fe alloys for various Fe concentrations (after Ref. [49]).

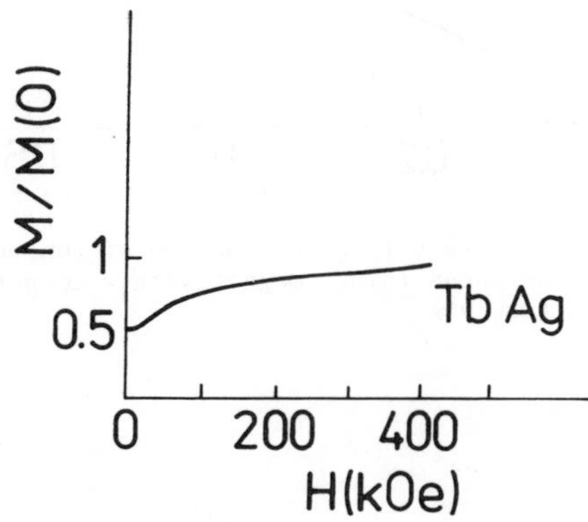

Fig.12. Magnetization of amorphous TbAg alloy in high fields (after Ref. [49]).

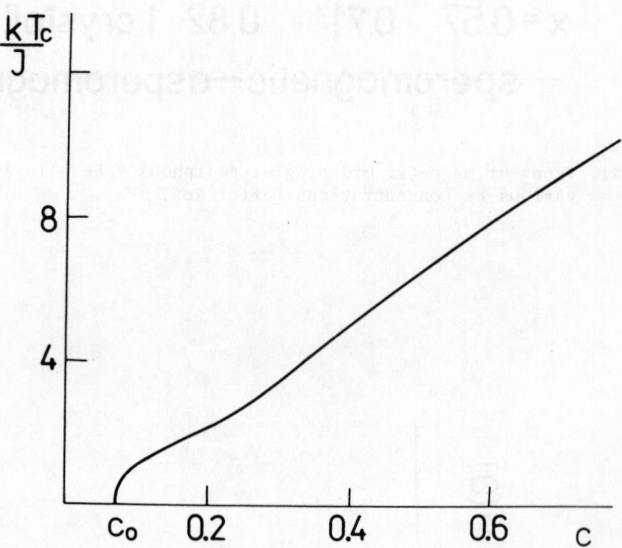

Fig.13. Critical temperature T_c as a function of concentration c for a diluted amorphous Ising magnet with a computer-simulated DRPHS structure [64].

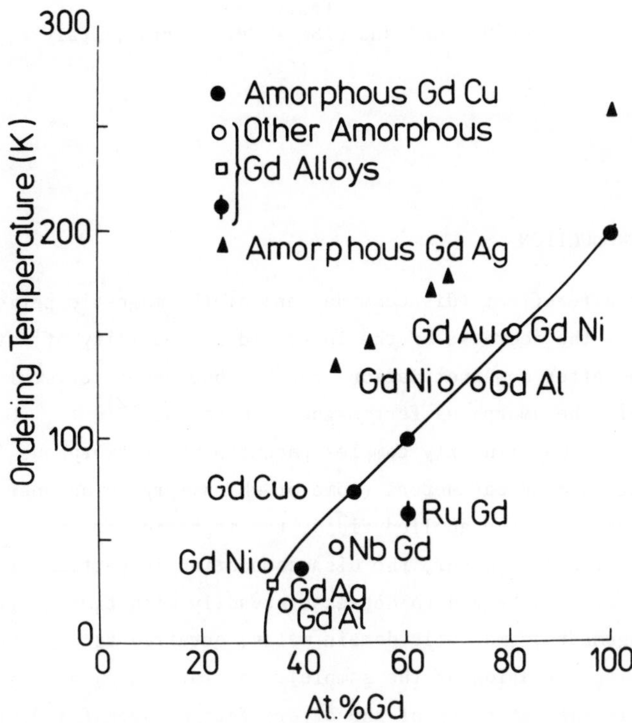

Fig.14. Magnetic ordering temperature for amorphous Gd alloys. The curve represents a fit to the Bethe-Peierls-Weiss approximation formula $J_o/k_BT_c = \frac{1}{2}\ln\frac{cz}{cz-4}$ and z = 12. Data taken from Refs [1,72,73,93].

MAGNETIC AFTEREFFECTS IN AMORPHOUS FERROMAGNETIC ALLOYS

P. Allia and F. Vinai
Istituto Elettrotecnico Nazionale "G. Ferraris", Torino
Italy
GNSM - CNR and CISM - MPI, Torino, Italy

1. INTRODUCTION

The aftereffect (disaccomodation) of the magnetic permeability μ_i, defined as the decrease of the low-field permeability of a ferromagnet with time after demagnetization, can be observed in crystalline [1)] as well as in the amorphous ferromagnetic materials [2)].

It is a particularly complex phenomenon, in amorphous ferromagnets depending on many parameters (some of which largely uncontrolled) as the intensity of the applied field, the magnetic domain structure, the internal stress. However, the disaccomodation is particulary easy to measure: μ_i is observed to decrease steadily with time t after any re-arrangement of magnetic domain walls, occurred at a given t_0 (e.g., after demagnetization of the sample). The ratio $\Delta\mu/\mu = [\mu_i(t_1)-\mu_i(t_2)]/\mu_i(t_2)$ is the intensity of the aftereffect between fixed times ($t_1 > t_0$; $t_2 > t_1 > t_0$), which are determined by experimental conditions.

Experimental methods allowing one to explore the dynamics of the permeability decay in an extremely wide time interval ($\simeq 10^{-5}$s to $\simeq 10^5$s after demagnetization) are now currently available. In strong contrast to the fast development of experimental methods, the present state of theories on the nature of the processes giving rise to the magnetic aftereffect in amorphous alloys is much less advanced, although recently, a particularly good description of its complex character has been published [3].

Generally, the permability aftereffect is related to the progressive decrease of mobility of 180° Bloch walls submitted to a low a.c. driving field and reversibly moving about their equilibrium positions. $\Delta\mu/\mu$ originates from directional ordering of systems interacting with magnetization, and being oriented by means of thermally activated motion, to follow the changes of the magnetization direction. The magnetic disaccomodation observed in soft ferromagnets is thus entirely related to the changes of the magnetization direction occurring within 180° Bloch walls during their motion.

Very strong aftereffects are usually detected in amorphous ferromagnets. In contrast to the ferromagnetic crystals (where the magnetic aftereffects are undoubtedly related to the presence of specific solute atoms (or atomic pairs) whose ordering is characterized by rather sharp activation energies[1]) in the case of amorphous ferromagnets the collected data strongly support the hypothesis of the existence of a broad, structureless distribution of activation energies for the aftereffect. Great differences exist, in fact, between the magnetic disaccomodation phenomena observed in the crystals, and those occurring in the amorphous metals:

- the time behaviour of the initial permeability after demagnetization in crystalline alloys is typically $\mu_i(t) = A+Be^{-t/\tau_o}$ (where τ_o is a single relaxation time); in amorphous systems $\mu_i(t)$ is generally well approximated by the law of the type $a-b\cdot\ln t$;
- the temperature behaviour is completely different. In crystalline ferromagnetic alloys the aftereffect of the magnetic permeability can only be observed in a very narrow temperature range, since the relaxation time τ_o, related to the activation energy Q_o by a conventional Arrhenius relation, is estremely sensitive to any relevant change of temperature. However, in amorphous ferromagnets the magnetic aftereffect can be observed at any temperature between 4 K and the Curie point for the alloy [2].

The intensity of relaxation is temperature dependent, generally increasing with the increase of T. From the study of the behaviour of

$\Delta\mu/\mu$ vs. temperature, it is possible, under certain assumptions, to derive an approximate form of the distribution p (Q) of activation energies for the aftereffect, which appears, indeed, to be essentially structureless and completely independent of the actual alloy composition [2].

Such results have been interpreted in terms of some models [4,5,6], whose common point is the consideration of the intrinsic effects of local disorder on the microscopic ordering processes giving rise to the aftereffect. Different results confirm this viewpoint. (i) The disaccomodation completely disappears after crystallization of the material, even if soft ferromagnetic crystalline phases are still present. (ii) Any structural relaxation involving irreversible changes of topological short-range order (TSRO) induces irreversible changes of $\Delta\mu/\mu$ [7]. (iii) Several measurements have shown that the maximum intensity of the aftereffect measured for as-quenched samples at constant temperature is strongly dependent on the quenching rate at which the amorphous ribbons vere produced. In particular, ribbons prepared with higher quenching rates (thus having larger frozen-in free volume, and greater structural disorder) display higher $\Delta\mu/\mu$ values [8]. This is a general rule, tested in different sets of samples, all prepared with special care to avoid any appreciable shift in composition for samples belonging to the same "family".

These data indicate that the degree of amorphicity of a glassy metal plays a relevant role in determining the intensity of the aftereffect at a given temperature. On the other hand, few attempts have been made to explicitly define and characterize the microscopic ordering processes which are the source of magnetic disaccomodation. In certain cases, in fact, simple models for these processes have been introduced [6,7]. They were based on rather reasonable assumptions, but did not lead to any quantitative theory capable to explain most of the data with a few simple assumptions. Our feeling is that this is indeed possible.

A detailed explanation of the concepts used in a new structural approach to the problem [9], will be given in Section 3.

2. MEASURING METHODS

In principle the disaccomodation measurements are not difficult. After sample demagnetization, the behaviour with time of the a.c. magnetic permeability is followed in time interval whose limits (t_1 and t_2) are controlled (e.g., by a computer) and kept fixed in order to allow one to make a comparison among different permeability decays. Obviously, the frequency of the driving field must be low enough so that the dissipative effects could be neglected. A measurement is actually an average over various permeability decays, in order to minimize the experimental fluctuations, which are rather large in this case.

Obviously the main experimental problem with amorphous alloys arises from the quasi-logarithmic behaviour of $\mu_i(t)$ after demagnetization. In this case, in fact, a considerable part of the phenomenon takes place in very short time intervals ($t \lesssim 1$ s). However the conventional demagnetization methods (e.c, a.c. demagnetization at $\nu = 50$ Hz) do not allow to study times shorter than 1 s. Our group succeeded in developing an impulsive method allowing much short times to be studied. In this case the demagnetization is substituted by a sudden change of the magnetic domain pattern of the sample at remanence, obtained by superposing to a small driving field a disordering square-wave field of intensity close to the material's coercivity.

In this way, the behaviour of $\mu_i(t)$ may be followed from about 10^{-2}s after each domain pattern change [2].

Even shorter times (10^{-5}s) may be investigated if using the same technique but recently improved, described in detail elsewhere [10].

3. STRUCTURAL MODEL FOR THE AFTEREFFECT

Our starting point is the recognition that the intensity of aftereffect at room temperature is proportional to the square of the saturation magnetostriction constant λ_s measured at the same temperature [4]. This is certainly true, at least in magnetostrictive Fe-based alloys $1 \cdot 10^{-6} < \lambda_s < 40 \cdot 10^{-6}$; cf. Fig.1. As a consequence, it is reasonable to assume the existence of magnetostrictive coupling between the ordering defects present in amorphous metals and the magnetization. In this approach, the defects are best characterized in terms of local stresses and strains. Thus it is quite natural to adopt the concepts introduced by T. Egami and coworkers [11,12], who described structural properties of a glassy metal model in terms of the fluctuations of local (hydrostatic and shear) stresses. These stresses can be shown to be proportional to local strains, describing the degree of structural distortion of the nearest neighbour shell of any atom in the material [12]. The Cartesian, symmetric strain tensor defined at a given site is described by six independent components. It is possible to introduce a symmetry representation of the tensor, in which one component transforms under rotation of the reference system like a spherical harmonic function with l=0, the remaining five components transforming as the five spherical harmonics with l=2 ($-2 < m < 2$). In this way, it is possible to naturally split the strain tensor in one isotropic and five anisotropic component which can be written, in terms of the conventional Cartesian components, as

$$
\begin{aligned}
\varepsilon_a &= \sqrt{1/3} \cdot (\varepsilon_{11}+\varepsilon_{22}+\varepsilon_{33}) \\
\varepsilon_{\gamma I} &= \sqrt{2/3} \cdot (\varepsilon_{33}-(\varepsilon_{11}+\varepsilon_{22})/2) \\
\varepsilon_{\gamma 2} &= \sqrt{1/2} \cdot (\varepsilon_{11}-\varepsilon_{22}) \; ; \; \varepsilon_{\varepsilon I} = \sqrt{1/2}\, \varepsilon_{23} \\
\varepsilon_{\varepsilon 2} &= \sqrt{1/2}\, \varepsilon_{12} \; ; \quad \varepsilon_{\varepsilon 3} = \sqrt{1/2}\, \varepsilon_{31}
\end{aligned} \tag{1}
$$

where the subscripts $\mu = (\gamma_I, \gamma_2, \varepsilon_I, \varepsilon_2, \varepsilon_3,)$ label pure shear strains, ε_a being a pure compression/dilation. From now on, the subscript μ will stand for the set of shear strains only, with the exclusion of the hydrostatic strain. Generally, a site where at least one of the ε_μ's is high, is strongly sheared, i.e. the nearest neighbour shell is severely distorted. It should be noted that the anisotropic components of the strain tensor defined in Eq. (1) transform like Tesseral harmonics $Z_{2\mu}$, i.e. linear combinations of the five spherical harmonics Y^m_2. Tesseral harmonics are preferred for our purposes because they are real functions.

The existence of strong strains frozen in ferromagnetic amorphous material, together with the presence of anisotropic magnetic interactions between constituent atoms, leads one to consider a site-dependent magneto-elastic energy F_{me}, representing the effect of the local strain on the magnetic anisotropy energy F_a.

This structural magneto-elastic energy can be shown[9)] to have the form (to the first order in the local shear-stresses σ_μ):

$$F_{me}(i) = \Sigma_{\mu\mu'}(\lambda_c\delta_{\mu\mu'}+\lambda_{\mu\mu'})\sigma_{\mu'}\cdot Z_{2\mu}(\Omega_M) \tag{2}$$

where Ω_M is the magnetization direction, in polar coordinates, in a local arbitrary reference system, and σ_μ is proportional to ε_μ by the average shear elastic constant $\bar{C}_\gamma$ [12)]. Finally,

$$\lambda_c = -\sqrt{8\pi/15}\,(\bar{z}/2)\,\bar{C}_\gamma^{-1}\cdot f(a) \tag{3a}$$

$$\lambda_{\mu\mu'} = -\sqrt{8\pi/15}\cdot(4\pi/5)\,\bar{C}_\gamma^{-1}((1/2)\,af'(a)-f(a))\cdot$$

$$\cdot\,\Sigma_j\,Z_{2\mu}(\Omega_{ij})Z_{2\mu'}(\Omega_{ij}) \tag{3b}$$

where $\bar{z}$ is the average coordination number, "a" is an average nearest-neighbour distance, f(r) is the radial part of the local anisotropy energy, and the summation is performed over neighbours of the i-th atom.

While λ_C is homogeneous in the material, $\lambda_{\mu\mu'}$ is site-dependent and involves information about the symmetry of the bond directions Ω_{ij} of the neighbours of the reference i-th atom. Since f(r) is a very rapidly decreasing function of r, the dominant contribution to the summation in Eq. (3b) essentially comes from atoms within the first peak of the material's RDF.

We have shown [9)] that an expression for the magnetic aftereffect of magnetostrictive nature may be derived from Eq. (2) in the form:

$$\frac{\Delta\mu}{\mu} = 2.55\,\frac{.3}{32}\,\frac{N_T}{kT}\,\frac{G(t_1)-G(t_2)}{H_e I_s}\,\lambda^2_{eff}\,\langle\tau^2\rangle \tag{4}$$

where N_T is the number of defects moving at a given temperature T, H_e - the applied field, I_s - the saturation magnetization, G(t) - a decreasing function of time, $\langle\tau^2\rangle$ is the second moment of the shear-stress fluctuations, and λ^2_{eff} is defined as:

$$\lambda^2_{eff} = (8\pi/25)\cdot(\lambda^2_s + <m> \lambda^2_a) \quad (5)$$

λ_s being the saturation magnetostriction, and $\lambda_a = -(1/4\pi)\cdot\sqrt{8\pi/15}\cdot\bar{z}\cdot\bar{C}^{-1}_{\gamma}[\frac{1}{2}a\, f'(a)-f(a)]$. Finally, $< m >$ is the average value of a local order parameter

$$m_i = \{\frac{5}{2z_i} \Sigma_{jk}[P_2(\cos\vartheta_{jk})]^2-1\} \quad (6)$$

the double summation being performed over the z_i nearest-neighbour bond directions around atom i, and the P_2's being second-order Legendre polynomials.

$\Delta\mu/\mu$ is thus related, for the first time, to a structural parameter m_i, whose properties make it very useful to describe the angular distortion of nearest neighbour shells, as discussed elsewhere [9].

The detailed information about the behaviour of $\Delta\mu/\mu$ can be obtained by analyzing the results of the present approach. It can be shown that λ_s and λ_a cannot vanish simultaneously [9]; as a consequence, λ_{eff} is always different from zero. In particular, when $\lambda_s = 0$ (zero-magnetostriction ribbons),

$$\Delta\mu/\mu \sim \lambda^2_{eff} \sim < m > \lambda^2_a \neq 0. \quad (7)$$

This result explains why magnetic aftereffects, although of strictly magnetostrictive nature, can be observed also in alloys with vanishingly small macroscopic magnetostriction.

Finally, it is suggested that the dramatic changes induced in $\Delta\mu/\mu$ by varying the amorphicity conditions of the material [8], may be at least partly related to the presence of the factor $< m >$ in λ^2_{eff}, which has been shown to be particularly sensitive to the structural disorder, and is lowered by structural relaxation, as a consequence of the atomic re-arrangement during this process.

4. OTHER APPROACHES

Recently, Kronmuller et al. [6,13] examined the possible sources of the coupling between structural defects and magnetization. Starting from analysis of micromagnetics, they concluded that this interaction consists of spin-orbit and exchange contributions, in addition to the magneto-elastic term first depicted by Allia and Vinai. It is indeed likely that additional terms are present in the coupling energy responsible for magnetic aftereffect. However, our viewpoint is that the only way to give a good answer to the problem is to perform new experiments, rather than to make a comparison between models based on different assumptions. Unfortunately, a critical experiment indicating which type of energy plays a predominant role in the considered processes, has not yet been performed. It is worth mentioning that historical development of the concepts concerning magnetic aftereffects in crystalline ferromagnets followed a closely similar path [14]. In fact two different theories were proposed to explain the magnetic aftereffects due to C atoms in α-Fe: the first, based on the magneto-elastic picture, by Snoek; the second one, giving emphasis to the local anisotropy energy, by Neel. An illuminating experiment performed by De Vries et al. showed in that case the anisotropy contribution was dominating over the magneto-elastic one.

This conclusion cannot be directly extrapolated to the present case, since the structural distortions present in amorphous metals are much higher than the deformation locally induced by C interstitials in the α-Fe lattice, consequently increasing the importance of the magneto-elastic term, which seems to be larger than both the spin-orbit and the exchange terms, at least in fairly magnetostrictive alloys, according to Kronmuller's estimate [13].

REFERENCES

1) Slonczewski, J.C., in: Magnetism I, eds. Rado, G., Suhl, H.; p.205 (Academic, New York 1963).
2) Allia, P., Mazzetti, P., Soardo, G.P., Vinai, F., J. Magn. Magn. Mat. 19, 281 (1980).
3) Kronmuller, H., Phil. Mag. B 48, 127 (1983).
4) Allia, P., Vinai, F., Phys. Rev. B 26, 6141 (1982).

5) Kisdi-Koszo, E., Vojtanik, P., Potocky, L., J. Magn. Magn. Mat. 19, 159 (1980).

6) Kronmuller, H., Moser, N., Rettenmeier, F., IEEE Trans. Magnetics 20, 1388 (1984).

7) Allia, P., Sato Turtelli, R., Vinai, F., J. Magn. Magn. Mat. 39, 279 (1983).

8) Allia, P., Luborsky, F.E., Sato Turtelli, R., Soardo, G.P., Vinai, F., IEEE Trans. Magnetics 17, 2615 (1981).

9) Allia, P., Vinai, F., Phys. Rev. B 33, 422 (1986).

10) Allia, P., Beatrice, C., Vinai, F., J. Magn. Magn. Mat., in press.

11) Egami, T., Maeda, K., Vitek, V., Phil. Mag. A 41, 883 (1980).

12) Egami, T., Srolovitz, D., J. Phys. F 12, 2141 (1982).

13) Kronmuller, H., Phys. Status Solidi B 127, 531 (1985).

14) Cullity, B.D.: "Introduction to Magnetic Materials", p.369 (Addison-Wesley, Reading 1972).

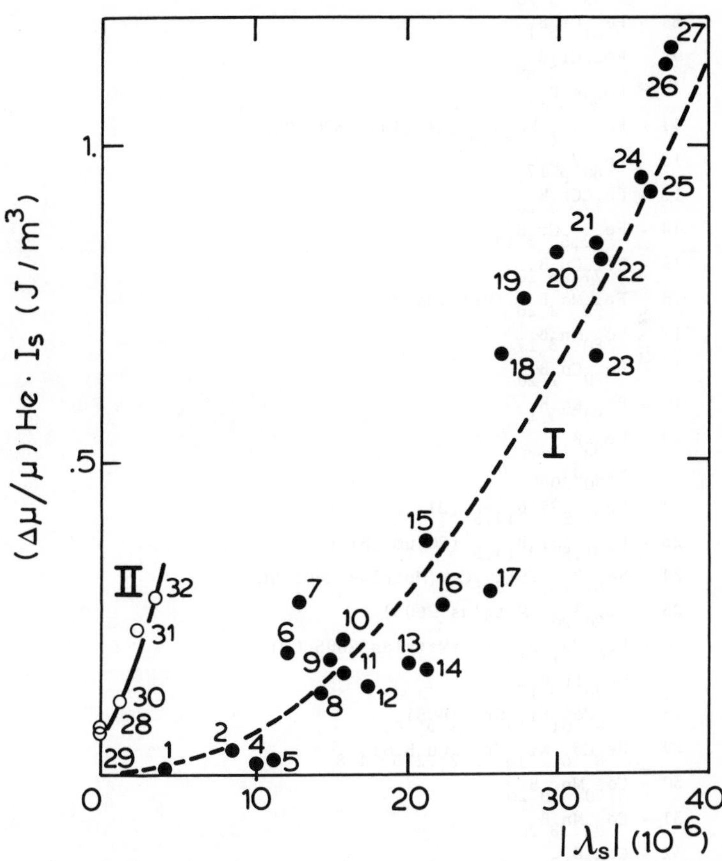

Fig.1. Magnetic disaccomodation(multiplied by the applied field times saturation magnetization [9]) vs. absolute value of saturation magnetostriction in Fe-based and Co-based alloys (curves I and II, respectively). Samples identified as follows:

1 - $Fe_{29}Ni_{49}P_{14}B_{6}Si_{2}$ (Metglas 2826 B)

2 - $Fe_{40}Ni_{40}P_{14}B_{6}$ (20 μm thick)

3 - $Fe_{40}Ni_{40}P_{14}B_{6}$ (30 μm thick)

4 - $Fe_{40}Ni_{40}P_{14}B_{6}$ (40 μm thick)

5 - $Fe_{40}Ni_{40}P_{14}B_{6}$ (Metglas 2826)

6 - $Fe_{80}Nb_{3}B_{17}$
7 - $Fe_{75}Cu_{5}B_{20}$
8 - $Fe_{81}Cr_{5}B_{14}$
9 - $Fe_{75}Cr_{5}B_{20}$
10 - $Fe_{80}W_{3}B_{17}$
11 - $Fe_{40}Ni_{38}Mo_{4}B_{18}$ (Metglas 2826 MB)
12 - $Fe_{80}V_{3}B_{17}$
13 - $Fe_{77}Cr_{3}B_{20}$
14 - $Fe_{82.5}Cr_{3}B_{14.5}$
15 - $Fe_{77}Cu_{3}B_{20}$
16 - $Fe_{78}Mo_{2}B_{20}$ (Metglas 2605A)
17 - $Fe_{80}Mn_{3}B_{17}$
18 - $Fe_{79}Cu_{1}B_{20}$
19 - $Fe_{80}Rh_{3}B_{17}$
20 - $Fe_{85}B_{15}$
21 - $Fe_{80}B_{20}$
22 - $Fe_{81.5}Si_{4}B_{14.5}$ (21 μm thick)
23 - $Fe_{81.5}Si_{4}B_{14.5}$ (26 μm thick)
24 - $Fe_{81}B_{13.5}Si_{3.5}C_{2}$ (Metglas 2605 SC)
25 - $Fe_{80}B_{20}$ (Metglas 2605)
26 - $Fe_{67}Co_{18}B_{14}Si_{1}$ (Metglas 2605 CO)
27 - $Fe_{80}Ir_{3}B_{17}$
28 - $Fe_{8}Co_{61}Ni_{19}Cr_{4.2}B_{3}Si_{4.8}$
29 - $Fe_{8}Co_{61}Ni_{18}Cr_{3.2}Cu_{2}B_{3}Si_{4.8}$
30 - $Co_{70}Mn_{1o}B_{2o}$
31 - $Co_{72}Mn_{8}B_{20}$
32 - $Co_{73}B_{17}$

RANDOM ANISOTROPY IN AMORPHOUS MAGNETIC ALLOYS

Henryk Szymczak

Institute of Physics, Polish Academy of Sciences
02-668 Warszawa, Al.Lotników 32/46, Poland

1. INTRODUCTION

Topological disorder in amorphous magnetic alloys leads to randomness in the interactions of the magnetic moments. Two of the most important kinds of interactions are usually considered:

- random isotropic exchange
- random anisotropy (single-ion or two-ion anisotropy).

Magnetic properties of amorphous alloys are governed, as in the case of their crystalline counterparts, by the interplay of exchange and anisotropic interactions. One of the main differences is that anisotropic interactions in crystals tend to align the magnetic moments along a few equivalent crystallographic directions whereas in amorphous alloys that cannot be the case.

The first theoretical descriptions [1,2] of magnetism in amorphous solids were based on the assumption that the most important effect of amorphous structure on magnetism is that it introduces fluctuations in the exchange interactions. However, it is commonly accepted that a reasonable amount of randomness in the isotropic exchange is not expected to qualitatively affect the magnetic behaviour of solids, even near the critical point. Random anisotropy, on the other hand, which is present in all amorphous materials, can dramatically affect the magnetic behaviour.

The effect of random anisotropy was first observed in the low temperature properties (magnetization curve, coercive field) of amorphous rare earth-transition metal alloys by Clark [3]. The observations of this unusual ferromagnetic behaviour of amorphous intermetallic compounds led Harris, Plischke and Zuckermann [4] to propose that the magnetic properties of amorphous magnets would be strongly influenced by random local electrostatic fields. They assumed that each rare earth ion is subjected to a random magnetic anisotropy whose origin is responsible for the large magnetocrystalline anisotropy in rare earth crystalline alloys (single-ion mechanism).

This paper deals with the effect of random anisotropy on the following magnetic properties of amorphous ferromagnets:

a) magnetization curve

b) magnetostriction

c) magnetic ordering.

Other magnetic properties of random anisotropy systems are excellently reviewed by Moorjani and Coey [5] and by Kaneyoshi [6].

2. RANDOM MAGNETIC ANISOTROPY MODEL

The model introduced by Harris, Plischke and Zuckermann [4] is defined by the following Hamiltonian:

$$H = -\frac{1}{2}J\sum_{ij}\bar{S}_i\cdot\bar{S}_j - D\sum_i(\bar{n}_i\cdot\bar{S}_i)^2 - g\beta\bar{H}\cdot\sum_i\bar{S}_i \qquad (1)$$

Here $\bar{S}_i$ is the total angular momentum at site i; J_{ij} ($J_{ij} \geq 0$) are the ferromagnetic exchange interactions of any range; $\bar{n}_i$ denotes the direction of the local anisotropy, $D(D > 0)$, at site i, and H is a magnetic field.

In (1) β denotes the Bohr magneton and g is the g-factor of magnetic atoms.

It follows from eq. (1) that in the random anisotropy model the fluctuations in exchange are ignored but a site-dependent uniaxial anisotropy term is taken into account whose strength D remains constant but whose orientation ($\bar{n}_i$) varies randomly from site to site.

In order to make the Hamiltonian (1) tractable, the simplest way is to apply the molecular field approximation. In this case the Hamiltonian (1) is given by

$$H = - \sum_i [D(\bar{n}_i \bar{S}_i)^2 + \lambda \bar{S}_{iz}] \tag{2}$$

where

$$\lambda = pJ\langle\langle S_z \rangle\rangle + g\beta H \tag{3}$$

p is the average number of magnetic nearest neighbours and z is the external field direction;
<< ... >> implies both a thermal average, and an average over the random angles ϕ_i between the local easy axis $\bar{n}_i$ and the z-axis.

3. MAGNETIZATION CURVE [7)]

In the following we limit ourselves to the case T = 0 and to the classical case ($S \to \infty$). The reduced magnetization curve may be calculated by minimizing the classical energy ε $(H/pJ^2S \to \varepsilon)$

$$\varepsilon = - \sum_i [(h+M)\cos\theta_i - \alpha \cos^2(\phi_i - \theta_i)] \tag{4}$$

where θ_i is the angle between the z-axis and the spin direction,

$$h = g\beta H/(pJS), \qquad \alpha = D/(pJ) \qquad \text{and}$$

$$M = \langle \cos\theta_i \rangle = \frac{\int \sin\phi_i \cos[\theta_i(\phi_i)] d\phi_i}{\int \sin\phi_i \, d\phi_i} \tag{5}$$

By minimizing ε with respect to θ_i, the equilibrium position of the spin is obtained from

$$(h+M)\sin\theta_i = \alpha \sin 2(\phi_i - \theta_i) \tag{6}$$

Equations (5) and (6) form a set of self-consistent equations for the normalized spontaneous magnetization M at zero temperature. In order to obtain a magnetization curve M(h) one should solve the eqs. (5) and (6) numerically. In any case the magnetic structure in zero applied field will be asperomagnetic. The normalized spontaneous magnetization may be calculated directly in the following two cases:

a) small anisotropy $\alpha \ll 1$

$$M = 1 - \frac{4}{15}\alpha^2$$

b) high anisotropy $\alpha \gg 1$

$$M = \frac{1}{2}(1 + \frac{1}{3\alpha})$$

The decrease of M with α from the value M = 1 (for $\alpha = 0$) to the value M = 1/2 (for $\alpha = \infty$) is due to competition between the exchange interaction and the random magnetic anisotropy.

Eqs. (5) and (6) may be used in order to study intrinsic coercivity (i.e. coercivity which does not depend on domain structure). The basic idea is that in a finite magnetic field along the positive z-axis, each individual spin has two orientations of minimum energy; on the stable minimum and the other of higher energy (meta-stable). From numerical calculations it can be concluded that

for $\alpha < 0.6$ intrinsic coercivity $h_c = 0$

for $0.6 < \alpha < 5.3$ intrinsic coercivity $h_c = 1.087\alpha - 0.65$

for $\alpha > 5.3$ intrinsic coercivity $h_c = 0.964\alpha$

4. MAGNETOSTRICTION

The random anisotropy model of magnetostriction in amorphous alloys has been developed in [8,9]. The model is based on the cluster theory of neɛ anisotropy in amorphous ferromagnets [10], with the energy of the system given by eq. (2). The local magnetoelastic tensor B_{ijkl} has the form (written in Voigt notation) determined by axial symmetry:

$$[B] = \begin{vmatrix} B_{11} & B_{12} & B_{13} & 0 & 0 & 0 \\ B_{12} & B_{11} & B_{13} & 0 & 0 & 0 \\ B_{31} & B_{31} & B_{33} & 0 & 0 & 0 \\ 0 & 0 & 0 & B_{44} & 0 & 0 \\ 0 & 0 & 0 & 0 & B_{44} & 0 \\ 0 & 0 & 0 & 0 & 0 & B_{66} \end{vmatrix}$$

where $B_{31} = -(B_{11}+B_{12})$, $B_{33} = -2B_{13}$, $B_{66} = \frac{1}{2}(B_{11}-B_{12})$ and the z-axis is parallel to the local anisotropy axis. After averaging the local magnetoelastic energy (taking into account the condition (6)) the essential part of the effective magnetoelastic tensor B^{eff}_{ijkl} will have isotropic properties and will be determined by one component B^{eff}_{11}:

$$\begin{aligned} B^{eff}_{11} &= B_{11}\langle(\cos^2\phi_i - \tfrac{2}{3})\cos 2(\phi_i-\theta_i) + \tfrac{1}{3}\cos^2(\phi_i-\theta_i)\rangle \\ &+ B_{12}\langle(\cos^2\phi_i - \tfrac{2}{3})\cos^2(\phi_i-\theta_i) + \tfrac{1}{3}\cos 2(\phi_i-\theta_i)\rangle \\ &+ B_{13}\langle(\cos^2\phi_i - \tfrac{1}{3})(1-3\cos^2(\phi_i-\theta_i))\rangle \\ &+ B_{44}\langle\sin 2\phi_i \sin 2(\phi_i-\theta_i)\rangle \end{aligned} \tag{7}$$

In (7) the bracket $\langle\ldots\rangle$ denotes averaging over all the directions of local anisotropy.

The B^{eff}_{11} value depends on the value of the local magnetoelastic tensor components B_{kl} and simultaneously on the α value. Therefore the temperature and field dependence of magnetostriction in amorphous alloys has two different sources. These are the temperature and field dependent parts of local magnetoelastic tensor components $B_{ij}(T,H)$ and the intrinsic amorphous contribution resulting from the averaging procedure in (7). The averaging procedure depends strongly on the value $\alpha(\sim D/J)$. In this way amorphization leads to a specific temperature and field dependence of magnetoelastic tensor components, different from the predictions of the single-ion or two-ion theories developed for crystalline magnets.

5. MAGNETIC ORDERING

Aharoni and Pytte [11] have predicted that a random anisotropy system would have qualitatively different critical properties from those of ferromagnets. Particularly, they have shown that the system would undergo a second order phase transition at some temperature T_R, to a very unusual, new magnetic state with an infinite magnetic susceptibility in the small D/J case for $T < T_R$ but without spontaneous magnetization at any temperature. However, in their later work [12] it was concluded that the susceptibility is limited for $T < T_R$ to

$$\chi_{max} \sim (J/D)^4 \tag{8}$$

This result has been confirmed experimentally by Barbara and Dieny[13,14] who studied the critical properties of some amorphous rare earth magnets.

Recently, the theory of magnetic ordering in the random anisotropy system has been developed [15,16] in both small and large D/J limits. For large D/J values each magnetic moment is directed almost along the random anisotropy axis at its site. The magnetic susceptibility in this case is very small and only in a very large magnetic field does the reorientation of magnetic moments occur to the hemisphere defined by the field direction. This frozen, random magnetic structure, with no long--range order, is called a speromagnet in Coey's [17] terminology. This structure is similar to that of a spin glass.

A much more complicated situation occurs in the case of weak random anisotropy. In this case three different magnetic structures are predicted [16], according to the strength of the external magnetic field H. In a low field regime the system is in a correlated spin glass phase. It has a very large magnetic susceptibility and a net magnetization equal to zero. A relatively small magnetic field nearly aligns the correlated spin glass system, producing a new phase-ferromagnet with a wandering axis. Thus this new system has a slightly noncollinear structure in which the tipping of the magnetization with respect to the external field varies over the sample. The tipping angle is correlated over a correlation length which decreases as the magnetic field increases. In the third-high field regime the tripping angles are completely uncorrelated from site to site. Available experimental data [18] appear to be consistent with the theory.

The fact that amorphous materials described by the random anisotropy model can never be ferromagnetic may be explained easily by using the so called "domain argument" [19,20]. Suppose that in ferromagnetic medium owing to random fluctuations, there is a large region of linear size L within which a preponderance of local axes are tilted by some angle (for simplicity - by angle $\pi/2$). The total energy change due to the presence of domains with tilted local axes is roughly estimated to be

$$\Delta E \sim JL^{d-2} - DL^{d/2} \tag{8}$$

where d is the space dimensionality.

If $\Delta E>0$ for large L then the ferromagnetic ground state is stable against the tilting of magnetic moments in response to fluctuations of random axes. This will be true for $d > 4$. If $\Delta E < 0$, i.e. for $d < 4$ the magnetic moments will tilt, destroying the ferromagnetism. It means that "the lower critical dimension", i.e. the dimension of the space below which ferromagnetism cannot exist - is 4.

REFERENCES

1) Gubanov, A.I., Sov. Phys. Solid State 2, 468 (1960).
2) Handrich, K., phys. status solidi (b) 32, K55 (1969).
3) Clark, A.E., Appl. Phys. Lett. 23, 642 (1973).
4) Harris, R., Plischke, M. and Zuckermann, M.J., Phys. Rev. Lett. 31, 160 (1973).
5) Moorjani, K. and Coey, J.M.D., Magnetic Glasses, Elsevier, Amsterdam, 1984.
6) Kaneyoshi, T., Amorphous Magnetism, CRC Press, Boca Raton, 1984.
7) Cochrane, R.W., Harris, R. and Zuckermann, M.J., Physics Reports 48, 1 (1978).
8) Szymczak, H. and Zuberek, R., IEEE Trans. Magn. MAG-17, 2843 (1981).
9) Szymczak, H. and Zuberek, R., J. Phys. F12, 1841 (1982).
10) Richards, P.M., AIP Conf. Proc. 29, 180 (1975).
11) Aharony, A. and Pytte, E., Phys. Rev. Lett. 45, 1583 (1980).
12) Aharony, A. and Pytte, E., Phys. Rev. B27, 5872 (1983).
13) Barbara, B. and Dieny, B., Physica 130B, 245 (1985).
14) Dieny, B. and Barbara, B., J. Physique 46, 293 (1985).
15) Chudnovsky, E.M. and Serota, R.A., J. Phys. C16, 4181 (1983).
16) Chudnovsky, E.M., Saslow, W.M. and Serota, R.A., Phys. Rev. B33, 251 (1986).
17) Coey, J.M.D., J. Appl. Phys. 49, 1646 (1978).
18) Sellmyer, D.J. and Nafis, S., J. Appl. Phys. 57, 3584 (1985).

19) Imry, Y. and Ma, S.-K., Phys. Rev. Lett. 35, 1399 (1975).
20) Pelcovitz, R.A., Pytte, E. and Rudnick, J., Phys. Rev. Lett. 40, 476 (1978).

MAGNETIZING PROCESSES IN METALLIC GLASSES

J. Fink-Finowicki, B. Lisowski

Institute of Physics, Polish Academy of Sciences
Al.Lotników 32/46, 02-668 Warsaw, Poland

1. INTRODUCTION

Magnetizing processes in amorphous alloys are of great technological importance as well as of general scientific interest. Excellent soft magnetic properties of these materials have evoked great interest in their technical application as a substitute for silicon-iron or other classical materials. Amorphous ribbons containing Fe, Ni or Co as metals and B, Si or P as metalloids are now commercially produced. They exhibit lower losses and higher magnetic permeability than the nowadays used oriented 3% silicon-iron sheets. The desirable magnetic properties can be achieved by appropriate choice of composition of the amorphous alloy. For example, it is possible to obtain the materials with large positive magnetostriction constant ($\lambda_s \simeq 40\times10^{-6}$ for FeSiB alloy) as well as the materials with the nearly-zero or negative magnetostriction for high Co contents.

Magnetic properties of amorphous alloys can be modified in very wide range by appropriate annealing in external magnetic field or under tensile stress. Investigations of magnetizing processes in metallic glasses prior to and after annealing are interesting from theoretical point of view and are also very important for applications of these materials. This review presents various aspects of magnetizing processes in amorphous ribbons.

2. DOMAIN STRUCTURE

Domain structure in amorphous alloys depends on the distribution of local magnetic anisotropy. Therefore observations of the arrangement of magnetic domains are an excellent method for study of the microstructural inhomogeneities. Domain structure influences strongly the magnetizing processes and therefore the knowledge of this structure is very important for investigations of almost all magnetic properties.

2.1 Observation Techniques

Magnetic domains in amorphous ribbons can be observed using scanning electron microscope or by means of the magnetooptical Kerr effect. The classical Bitter technique can be useful, but it gives exclusively the information about statical domain wall positions. Scanning electron microscope gives very good image of domains with high contrast and resolution. Kerr technique is very popular, but usually this method requires special polishing and electropolishing of samples. In order to increase the optical contrast the dielectric films of 1/4 wavelength thickness are evaporated on the sample surfaces. The surface preparation, which may change the original structure of sample, is the weak point of this method. In order to eliminate this disadvantage a sophisticated electronic image analysis can be used. This technique has been described recently by Schmidt et al. [1]. They achieved a significant improvement of domain contrast using a digital image subtraction and noise reduction so it was possible to observe the domain pattern on non-polished and non-coated amorphous ribbon surfaces.

Kerr effect was also successfully used for studies of local magnetization processes on the non-polished ribbon surfaces [2]. Focusing the laser beam on a small ($\phi \leqslant 10\ \mu m$) spot on the ribbon surface and using Wollanston prism for differential detection made it possible to analyse the reflected light polarization as a function of magnetic field. From such local hysteresis loop measurements the local magnetic anisotropy may be evaluated and the character of magnetizing processes may be determined.

Domain wall motion can be investigated using Kerr microscope with the light strobed in synchronization to the magnetizing frequency [3]. Using this technique it is possible to determine if the domain wall pattern is reproducible or not during magnetizing process. The simultaneous observation of domains on both ribbon surfaces is also possible using Kerr effect as it is reported by Salzmann et. al.[4].

2.2 Types of Domain Patterns

Although the domain pattern observed in amorphous ribbons is usually very irregular, two types of domains may be distinguished:

i) Wide domains with magnetization vector lying within the ribbon plane. Typical width of these domains is in the range of 100÷300μm. Domain pattern may be in a form of more or less regular stripes oriented parallel or perpendicular to the ribbon lehgth (Fig.1a) or it may appear to radiate out from a point (star-like domains - Fig. 1b).

ii) Narrow domains with magnetization perpendicular to the ribbon plane. Their typical width is about 1÷30μm In this case a closure domain structure is developed in order to avoid magnetic surface charges (see Fig.2).

Because of the lack of crystalline anisotropy, domain pattern of amorphous alloys is influenced by other additional anisotropies:

i) The shape anisotropy due to stray field of the specimen.

ii) The magnetoelastic energy due to the internal or external stresses in the magnetostrictive materials.

iii) The structure anisotropy related to the short-range ordering or to the induced atom pair ordering.

The shape anisotropy prefers the magnetization alignment parallel to the ribbon length (Fig.3a). The same configuration can be generated by an external tensile stress in the materials with positive magnetostriction λ_s. For materials with $\lambda_s < 0$ the tensile stress produces transversal (Fig.3b) or perpendicular (Fig.3c) domain pattern. As it is shown schematically in Fig.3, the comprehensive stresses act in the opposite manner.

A spatial distribution of local tensile and comprehensive stresses resulting from the melt-spinning process can produce a patches of narrow laminar domains surrounded by star-like wide domain structure as it is shown in Fig.1b. The behaviour of domain pattern in magnetic field and under external tensile stress was studied by Kronmüller's group [5,6,7]. They have found that the narrow domains disappear at much higher magnetic field than the wide domains. The narrow domains transform spontaneously into a laminar structure of wide domains at a definite critical value of tensile stress, both in the alloys with positive and negative magnetostriction. By an appropriate stress-relief annealing the number of narrow domain patches can be significantly reduced.

The easy axis of structure anisotropy in the as-quenched ribbons is usually aligned parallel to the flow direction of the melt. During the annealing process in an external magnetic field the structure anisotropy with the easy axis parallel to the annealing field is induced. In this way a desirable (longitudinal, transversal, perpendicular or oblique) direction of magnetic anisotropy may be achieved. Effect of annealing under stress is similar, but mechanically induced magnetic anisotropies appear even if annealing is performed above Curie temperature.

3. PRIMARY MAGNETIZING CURVE AND STATICAL HYSTERESIS LOOP

The shape of magnetizing curve depends on the behaviour of domain structure in external magnetic field. Let us consider the magnetizing processes for different domain patterns in magnetic field oriented parallel to the ribbon length. In the case of longitudinal magnetic anisotropy the shift of domain walls proceeds as it is shown schematically in Fig.4a. In the transverse domain configuration (see Fig.4b) the domain walls are stable and magnetic vectors inside domains rotate toward external field direction. For perpendicular domain structure in increasing magnetic field the domain wall displacements dominate at lower magnetic fields (Fig.5a) and at higher fields rotational processes occur (see Fig.5b).

In real cases, due to irregular domain structure both types of magnetizing processes can appear simultaneously. On primary magnetizing curve (Fig.6) four regions can be distinguished, in which different magnetizing processes are dominant:

i) reversible shifts of wide domain walls,
ii) irreversible shifts of wide domain walls,
iii) magnetization processes inside narrow domains,
iv) reversible rotation in a saturated sample.

Rayleigh region. In the low field region, which is extended up to $H = 0.4\ H_c$, the Rayleigh law is fulfilled:

$$M(H) = \chi_o H + \alpha H^2$$

where χ_o is the initial susceptibility and α is the Rayleigh constant. It is confirmed that χ_o, α and coercive field H_c for Fe-Ni based amorphous alloys are related by following relation[9]:

$$\frac{\alpha H_c}{\chi_o} = \frac{8}{\mu_o 3\sqrt{\pi}} \ln\left(\frac{D_o}{\lambda_o}\right)^{1/2}$$

where λ_o describes the average wavelength of the domain wall potential ($\sim 1\mu m$) and D_o is the width of the domains. The self-consistency of this relation with the experimental results for different $Fe_{80-x}Ni_xB_{20}$ alloys is shown in Fig.7.

Barkhausen noise. At higher fields the domain wall displacements become irreversible. Due to the large Barkhausen jumps the hysteresis loops are often irregular and magnetic noises in this region are observed. These fluctuation noises have been examined by Shirae [10] in different amorphous alloys, namely FeSiB, NiFeSiB and CoFeSiB. He has found the minimum value of noise for nearly-zero magnetostrictive $Co_{67}Fe_3Si_{15}B_{15}$ alloy (equal to one hundredth of the maximum value observed in Ni-based alloys). The magnetic noise can be significantly lowered by appropriate annealing process. The best results have been obtained by water quenching from annealing temperature. After such treatment the domain pattern becomes finer and in external magnetic field the number of domain walls moving simultaneously is larger, and this causes lower noise level. Shirae's conclusion is that some Co-based amorphous alloys have become comparable or superior to the conventional polycrystalline material such as Superpermalloy and can successfully be used in magnetic sensors of high sensitivity.

Propagation of irreversible magnetizing processes in amorphous ribbons was studied by Jansen et al.[11]. They have investigated the random Barkhausen jumps in $Fe_{78}Si_8B_{14}$ and $Fe_{40}Ni_{40}P_{14}B_6$ alloys and concluded that the propagation of magnetizing processes is controlled by eddy current damping.

The influence of external tensile stress on the magnetizing process in $Fe_{60}Ni_{20}B_{20}$ amorphous alloy was investigated by Filka et al.[12]. They have measured the field dependence of Barkhausen jumps dn/dH for different applied stresses (see Fig.8a). The dependence dn/dH versus H reaches its maximum for certain value of magnetic field, which is called the Barkhausen coercive field H_{cB}. The dependence of H_{cB} versus applied stress is shown in Fig.8b. The rapid increase of H_{cB} under stress of about 100 MPa is due to the process of the total reorganization of the domain structure (the vanishing of narrow domains). The presence of narrow laminar domains in the investigated samples have been confirmed by microscopic Kerr observations. The obtained results correspond to the influence of external stress on the domain structure, described in the previous chapter.

Saturation region. The high field region of magnetizing curve, with no domain structure, is characterized by the field dependence of the deviation of magnetization from its saturated value: $\Delta M(H) = M_s - M(H)$. As it was shown by Kronmüller[13], this deviation is caused predominantly by the structure defects of amorphous alloy and may be written as:

$$\Delta M = \frac{a_{1/2}}{H^{1/2}} + \frac{a_1}{H} + \frac{a_2}{H^2}$$

where each term of this equation may be attributed to special type of the stress source:

$a_{1/2}$ - to point-like defects with stress field σ varying as $1/r^3$,

a_1 - to quasi-dislocation dipoles with $\sigma \sim 1/r^2$.

a_2 - to isolated dislocations with $\sigma \sim 1/r$.

In amorphous ribbons the 1/H term dominates but in the magnetostrictionless Co-bassed alloys this term is a factor of 10 times smaller compared to the magnetostrictive Fe-Ni alloys.

Influence of annealing on hysteresis loop. The characteristic properties of magnetization process and of the hysteresis loop in the amorphous alloys are mainly determined by elastic stresses due to defect structures and by distribution of local magnetic anisotropies. Therefore the heat treatment, which can relieve internal stresses, strongly influences all magnetic properties of these materials.

Among a number of amorphous ferromagnets, there are a few alloys showing the Curie temperature T_c significantly lower than the crystallization temperature T_x. For these materials the considerable improvement of their soft magnetic properties can be achieved by simple non-field annealing at temperature above the Curie point T_c. Such treatment can relieve internal stresses and after annealing materials can exhibit very low coercive force and high permeability.

Quite different changes of magnetizing processes are observed after non-field annealing at temperature below T_c. During annealing the local anisotropy is induced inside the existing domain walls and therefore these walls tend to become pinning sites and nuclei of reverse domains. It leads to the domain wall stabilization and to the increase of coercive field. As an example, the changes of hysteresis loop after annealing in a demagnetized state for non-magnetostrictive Co-based ribbons are shown in Fig.9 after Yamasaki et al. [14]. Annealed ribbons exhibit the large Barkhausen jump and therefore they could be useful as Wiegand type magnetic sensors.

In order to relieve internal stresses but not to induce any uniaxial anisotropy, the annealing process can be performed in a rotating magnetic field as it was proposed by Kohmoto et al. [15]. This method may be useful especially for the amorphous materials with $T_c > T_x$, which cannot be annealed in the paramagnetic state.

Uniaxial magnetic anisotropy can be induced by annealing at $T < T_c$ in external magnetic field. Depending on orientation of the applied field, a desirable direction of this anisotropy can be obtained. The effect of different magnetic annealing treatments on the statical hysteresis loop for $Fe_{39}Ni_{39}Mo_4Si_6B_{12}$ alloy is shown in Fig. 10 after Warlimont [16]. After stress relief annealing in longitudinal magnetic field and slow cooling a nearly rectangular hysteresis loop is obtained with a small coercive field and an easy approach to magnetic saturation. Rapid cooling leads to a small value of induced anisotropy which results in a round hysteresis loop. A flat hysteresis loop results from annealing in a transverse magnetic field.

Whereas inducing the anisotropy by magnetic field annealing is effective only below the Curie temperature, a mechanical stress annealing below as well as above the T_c can induce the magnetic anisotropy. As it has recently been shown by Vazquez et al. [17] and by Nielsen et al. [18,19] changes of magnetic properties induced during annealing under tensile stress can be attributed to anelastic and plastic deformations. The anelastic deformations can be recovered by further non-stress annealing, whereas the plastic deformations are irreversible. It is interesting, that in some amorphous alloys the directions of induced anisotropy are different for these two types of deformation [19]

4. DYNAMIC PROPERTIES

The dynamic properties of magnetic materials are very important for most technical applications. Two aspects of this matter, strictly associated with magnetizing processes, are described in this chapter. One of them is the magnetic aftereffect and the other is the problem of total losses in amorphous magnetic alloys.

Magnetic aftereffect. Research of initial permeability and its disaccommodation has been motivated by the hope that studies of these magnetic properties may provide information on structural relaxation since they are very sensitive to microstructure. Disaccommodation of initial permeability is defined as:

$$D = \frac{\Delta\mu}{\mu} = \frac{\mu(t_2)-\mu(t_1)}{\mu(t_1)}$$

where μ is the value of initial permeability measured in ac magnetic field of a fixed frequency ω and amplitude H_a, t_1 and t_2 are the times of measurements after demagnetization of sample. The magnitude of the measuring ac field H_a is usually chosen as small as possible in order to be sure that the domain structure of the sample remains unchanged during the measurement. Similar formula can be used for definition of disaccommodation of imaginary permeability μ''. Sometimes the disaccommodation of the initial susceptibility $\chi(t)$ or of the reluctivity function $r(t) = 1/\chi(t)$ is used instead of disaccommodation of $\mu(t)$.

As it was shown by Allia et al. [20] the effect of disaccommodation in amorphous magnetics is mainly due to the reversible relaxation of structural defects. These defects are identified as high shear stress regions interacting via magnetostriction with the local magnetization inside the domain wall. Their theoretical model predicts that permeability aftereffect $\Delta\mu/\mu$ can be expressed by following equation:

$$\frac{\Delta\mu}{\mu} = A \frac{\lambda_s^2}{M_s} < \tau^2 >$$

where λ_s is saturation magnetostriction, A is a constant, $< \tau^2 >$ is the second moment of the shear stress fluctuation and M_s is saturation magnetization. Validity of this model has been confirmed for Fe-based alloys. The experimental values of $M_s \Delta\mu/\mu$ versus λ_s^2 in log-log scale, obtained for many different compositions (FeB, FeSiB, FeNiPB, FeCrB and FeCuB alloys) with λ_s between 1×10^{-6} and 40×10^{-6} lie on a straight line, in agreement with the proposed model. More detailed studies of magnetic aftereffect have been performed on series of FeCrCuB and FeCrB alloys after annealing. The observed influence of thermal treatment on the reduction of $\Delta\mu/\mu$ shows that disaccommodation must also be related to the macroscopic free volume [20]. It was also confirmed by the correlation between the temperature behaviour of magnetic aftereffect and of electrical resistivity [21].

Magnetic aftereffect of nearly non-magnetostrictive Co-based alloys as well as of magnetostrictive Fe-based alloys has been investigated by Kronmüller et al. [22]. They have measured the time and temperature dependence of initial susceptibility $\chi(t,T)$. The obtained results are shown in Fig.11 in the form of isochronal relaxation curves.

These spectra reveal three characteristic features:

i) relaxation nearly independent of temperature at low temperatures,

ii) a wide relaxation maximum in the temperature range from 250 to about 500 K,

iii) a second relaxation maximum, which could be detected in Co-based alloy only.

At the assumption of thermally activated processes with Arrhenius law this spectra have been analysed numerically and both the relaxation times and the distribution functions of the activation energies have been determined. From theoretical analysis based on the Anderson two level model, Kronmüller has concluded that relaxation effects in amorphous alloys are due to reorientation of atom pair axes with respect to the spontaneous magnetization within the domain wall. The effective interaction constant consists of the exchange energy and the local anisotropy energy as well as a magnetostrictive term, in contrast to the results of Allia and Vinai [23] who claimed that disaccommodation was caused exclusively by the magnetoelastic interactions.

Although disaccommodation is measured usually at very low ac magnetic fields, interesting phenomena can be observed at higher fields. As far as the amplitude of the measuring field is lower than the so-called stabilization field H_s, only the reversible wall movement should appear, but at higher fields irreversible jumps of the wall are expected. The measurements of aftereffect as a function of amplitude and frequency of ac magnetic field have been performed by Jagieliński et al. [24,25] on non-magnetostrictive FeCoSiB alloys. At external fields close to H_s they have observed very high values of disaccommodation in the investigated alloys (D_{10} measured at $t_2 = 10$ min after demagnetization reaches values up to 95%). Moreover, a sudden anomalous drop of permeability at certain characteristic time has been found in this material (see Fig.12a) [24]. It is suggested that this drop is due to rapid reduction of the number of domains. The field dependences of disaccommodation D_3 (for $t_2 = 3$ min) for as-quenched and annealed samples are shown in Fig.12b (after 25). The increase of low-field disaccommodation (at $H_a < H_s$) is related to the induced local anisotropies inside domains by annealing in demagnetized state. The stabilization field H_s is controlled by internal stresses produced by rapid quenching and reduction of this field after annealing is interpreted as a result of stress relief. For that reason the high-field aftereffect decreases during annealing.

Power losses. The total power loss P_T is a function of amplitude B_m and frequency f of induction. Measurements are usually carried out under sinusoidal flux condition [26]. Although the magnetic power losses are entirely generated by eddy currents generated by domain wall motions during magnetizing, they are traditionally divided into three major types, namely:

i) the static hysteresis loss P_h,
ii) the classical eddy curent loss P_e,
iii) the additional, so-called anomalous loss P_a.

The first term is proportional to the area of statical hysteresis loop and to the frequency f of alternating field. The experimental value of P_h can be extrapolated from the plot of total loss per cycle P_T/f against frequency f. In amorphous alloys this term is rather small due to the low value of static coercive field and it dominates only at low frequencies.

The second term, connected with eddy current surrounding the sample volume, is proportional to f^2, to the electrical conductivity and to the square of ribbon thickness. In amorphous alloys, due to small thickness and high resistivity, this term is lower than in crystalline alloys. Although P_e increases faster with frequency than P_h, typical value of classical eddy current loss in amorphous alloys does not exceed a few percent of total loss at 1 kHz frequency range [27].

The third term, i.e. anomalous power loss P_a dominates in amorphous alloys and is responsible for the non-linear behaviour of loss per cycle P_T/f versus frequency f. Traditionally P_a is compared to the classical eddy current loss P_e and the so-called loss anomaly factor η is defined as $\eta = 1 + P_a/P_e$. The typical values of η in metallic glasses vary from about 1000 to 100 in the frequency range 20 Hz ÷ 1 kHz and they can be reduced by one order of magnitude by appropriate annealing process [27].

The anomalous power loss originates from micro-eddy currents generated by the moving domain walls. Calculation of anomaly factor η for regular stripe domain pattern (longitudinal configuration) in magnetic sheet has been done by Pry and Bean [28] under assumption that domain walls move in a continuous and uniform way. Their results are shown in Fig.13 where the anomaly factor η for sinusoidal flux condition $b = B_m \sin\omega t$ is plotted as a function of the ratio of the domain width 2L to the sheet thickness d. The lower curve corresponds to the case $B_m/B_s \ll 1$ and the upper curve - $B_m/B_s = 1$ where B_s is saturation induction.

In the Pry and Bean model the anomaly factor does not depend on frequency. The non-linearity in the dependence of loss per cycle versus frequency can arise as a result of the spatial or temporal irregularities of domain wall motion or can be caused by the damping due to the viscosity field and relaxation processes. A general model of magnetic loss which takes account of these dissipative mechanisms has been proposed by Bertotti et al. [29].

Influence of the viscosity field on the power losses in amorphous alloys has been demonstrated by Celasco et al. [30]. Results of their measurements performed on $Fe_{40}Ni_{40}P_{14}B_6$ (Metglas 2826, are presented in Fig.14, where the dynamic power loss per cycle P/f is plotted versus magnetic induction derivative <dB/dt>. The upper curve corresponds to the as-obtained sample and two lower curves - to samples annealed at 480 K for 18h and at 673K for 6h, respectively. All measurements were performed at $B_m = 0.37T$. Classical eddy current loss is shown in an expanded scale. A small increase of domain width after annealing, detected in these samples using Kerr effect, cannot be responsible for the observed changes in power losses. The strong reduction of the loss after thermal annealing is attributed to reduction of the domain wall viscosity field due to stabilization of the amorphous atomic structure.

To obtain the desirable magnetic properties, amorphous robbons are usually annealed in a magnetic field oriented parallely to the ribbon axis. However after this type of annealing the amorphous glasses exhibit still large anomalous eddy current losses. In order to reduce these losses Fujimori et al. [31] have proposed the annealing in magnetic field tranverse or oblique to the ribbon axis. It produces lower ac losses because of the reduction of domain width and the increase of the rotational component in the magnetizing process. In the pure transversal geometry the magnetizing process consists in rotation exclusively, with no contribution from domain wall motion and thus there should be no anomalous losses.

As an example, the effect of the oblique field annealing on magnetic losses in amorphous $Fe_{78}Si_{10}B_{12}$ alloy is shown in Fig.15 after Kuo et al. [32]. The annealing processes have been performed at 400°C in magnetic field of 30 Oe applied at an angle θ to the ribbon axis. Fig. 15 shows the variation of the total loss per cycle P_T/f and also its constituent parts: P_h/f, P_e/f, and P_a/f for B_m = 1T, for two frequencies: 400Hz (Fig.15a) and 10 kHz (Fig.15b). In general, P_a decreases and P_h increases (due to increase of the coercive field) with the annealing field angle θ, while P_e remains constant at a given frequency. As a result, the optimum field direction for the lowest total loss is different for different frequencies. It is evident, that the oblique field annealing is an effective method for lowering the magnetic losses, especially for high frequency applications.

5. FINAL REMARKS

Still there is no any general theory which can describe properly all the aspects of magnetizing processes and their influence both on statical hysteresis loop and dynamical behaviour. This short review presented only the most important problems of magnetizing processes in metallic glasses. Special attention was paid to some practical aspects such as the relation between domain structure and magnetic properties and the influence of annealing treatment on magnetizing processes.

Bibliography contains only some of the papers in this field, chosen due to their importance and illustrative character.

REFERENCES

1) Schmidt, F., Schäfer, R., Hubert, A., Abstracts Soft Magn. Mat. 7, Blackpool, 1-6, Wolfson Centre for Magnetics Technology, Cardiff, (1985).

2) Dobrzański, A, Lisowski, B., Fink-Finowicki, J., Abstracts 2nd Int. Conf. on Magn. Mat., eds. Raułuszkiewicz, J. et al. 114, Jadwisin (1984).

3) Steck, G.J., Pregger, B.A., Kramer, J.J., J. Appl.Phys. 53, 7834 (1982).

4) Salzmann, P., Hubert, A., J. Magn. Magn. Mat. 24, 168 (1981).

5) Schroeder, G., Schäfer, R., Kronmüller, H., phys.stat.sol (a) 50, 475 (1978).

6) Dong, X.Z., Kronmüller, H., phys. stat. sol. (a) 70, 451 (1982)

7) Dong, X.Z., Gröger, B., Jendrysik, T., Kronmüller, H., phys. stat. sol. (a) 71, 441 (1982).

8) Kronmüller, H., Fähnle, M., Domann, M., Grimm, H., Grimm, R., Gröger, B., J. Magn.Magn.Mat. 13, 53 (1979).

9) Gröger, B., Kronmüller, H. J. Magn. Magn. Mat. 19, 161 (1980).

1o) Shirae, K., IEEE Trans.Magn. MAG-20, 1299 (1984).

11) Jansen, K., Grosse-Nobis, W., Kleibrink, H., J. Magn. Magn. Mat. 26, 267 (1982).

12) Filka, Š., Hajko, V., Zentko, A., Duhaj, P., Czechoslovak Jour. of Phys. B-33, 230 (1983).

13) Kronmüller, H., J. Appl. Phys. 52, 1859 (1981).

14) Yamasaki, J., Mohri, K., Watari, K., Narita, K, IEEE Trans. Magn. MAG-20, 1855 (1984).

15) Kohmoto, O., Fujishima, H., Ojima, T., IEEE Trans. Magn. MAG-1, 440 (1980).

16) Warlimont, H., Boll, R., J. Magn. Magn. Mat. 26, 97 (1982).

17) Vázquez, M., Fernengel, W., Kronmüller, H., phys. stat. sol.(a), 87, 609 (1985).

18) Nielsen, O.V., Hernando, A., Madurga, V., Gonzalez, J.M., J. Magn. Magn. Mat. 46, 341 (1985).

19) Nielsen, O.V., Barandiaran, J.M., Hernando, A., Madurga, V., J. Magn. Magn. Mat. 49, 124 (1985).

20) Allia, P., Soardo, G.P., Vinai, F., J. Magn. Magn. Mat. 31-34, 1527 (1983).

21) Allia, P., Andreone, D., Sato Turtelli, R., Vinai, F., Riontino, G., J. Magn. Magn. Mat. 26, 139 (1982).

22) Kronmüller, H., Moser, N., Rettenmeier, F., IEEE Trans. Magn. MAG-20, 1388 (1984).

23) Allia, P., Vinai, F., Phys. Rev. B-26, 6141 (1982).

24) Jagieliński, T., Walecki, T., Proc. of ICF III Int. Conf., Kyoto 652 (1980).

25) Jagieliński, T., J. Appl. Phys. 53, 2282 (1982).

26) Blundell, M., Overshott, K.J., Graham, C.D., Jr., J. Appl. Phys. 50, 1598 (1979).

27) Blundell, M.G., Graham, C.D., Jr., Overshott, K.J., J. Magn. Magn. Mat. 19, 174 (1980).

28) Pry, R.H., Bean, C.P., J. Appl. Phys. 29, 532 (1958).

29) Bertotti, G., Mazzetti, P., Soardo, G.P., J. Magn. Magn. Mat. 26, 225 (1982).

30) Celasco, M., Masoero, A., Mazzetti, P., Stepanescu, A., J. Magn. Magn. Mat. 31-34, 1407 (1983).

31) Fujimori, H., Yoshimoto, H., Morita, H., IEEE Trans. Magn. MAG-16, 1227 (1980).

32) Kuo, Y.C., Zhang, L.S., Gao, R.W., J. Magn. Magn. Mat. 31-34, 1563, (1983).

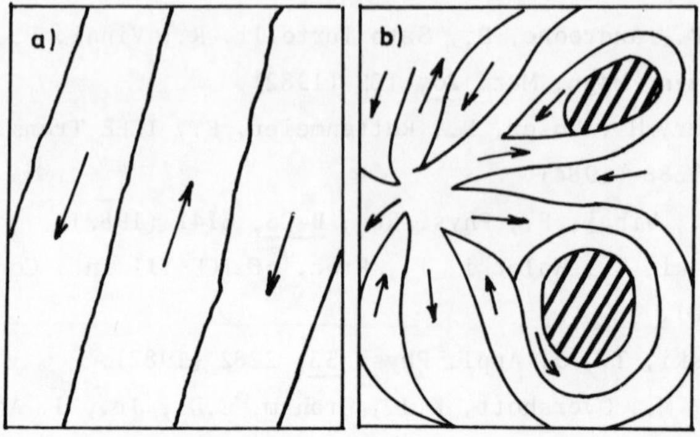

Fig.1. Typical domain patterns observed in amorphous ribbons: a) wide stripe domains, b) star-like structure with patches of narrow domains.

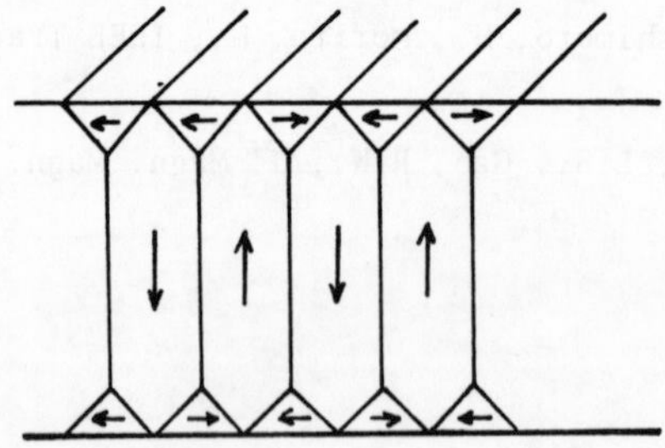

Fig.2. Model of narrow domain structure with closure domains.

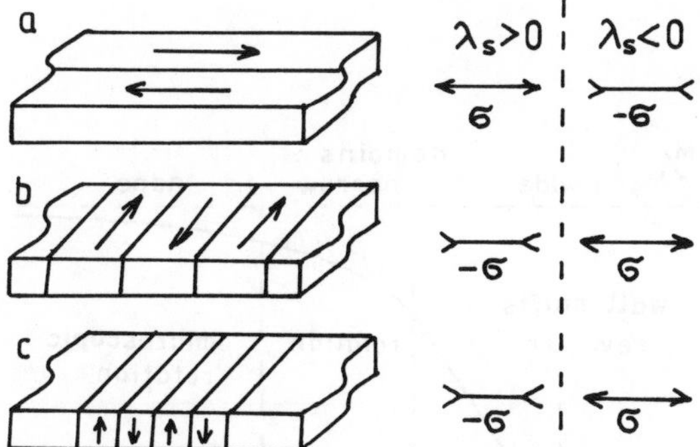

Fig.3. Types of domain structure of the materials with positive or negative magnetostriction under tensile stress: a) longitudinal, b) transversal and c) perpendicular structure.

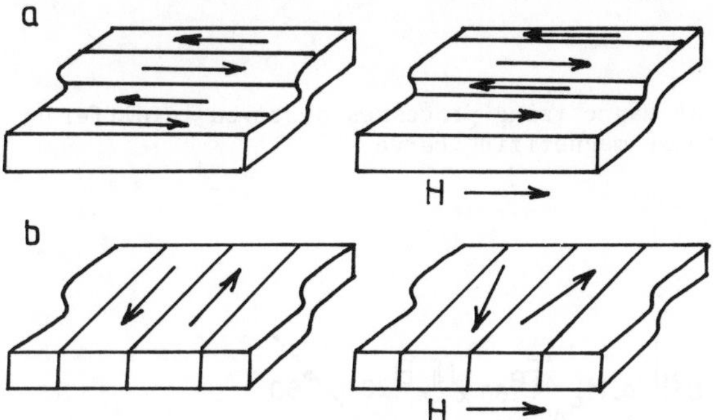

Fig.4. Magnetizing processes for a) longitudinal and b) transversal structure in an external magnetic field.

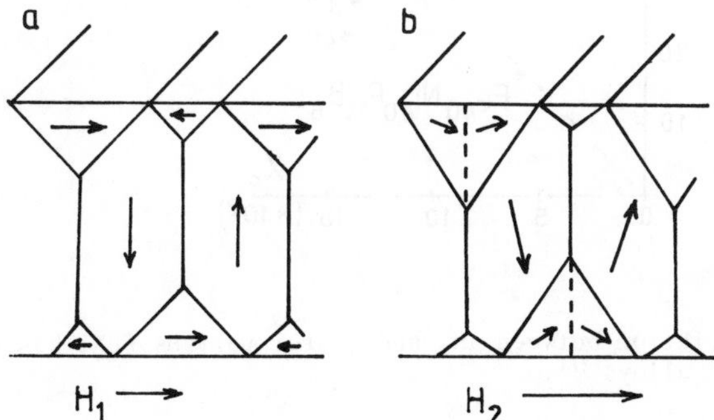

Fig.5. Perpendicular domain structure in an increasing magnetic field: a) shifts od domain walls, b) rotation of magnetization vector inside the domains [8].

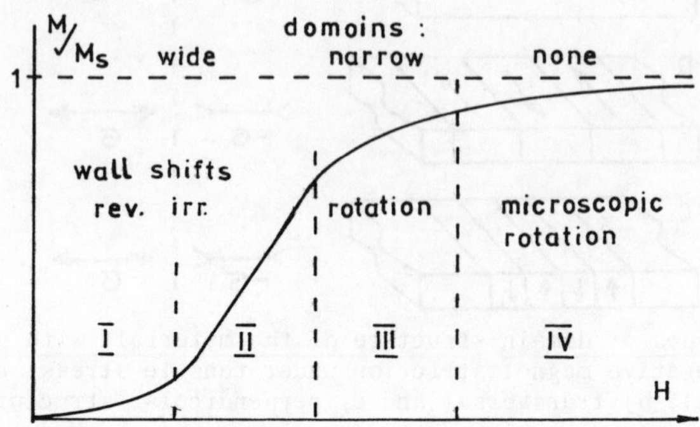

Fig.6. Types of magnetizing processes observed in different regions of primary magnetizing curve.

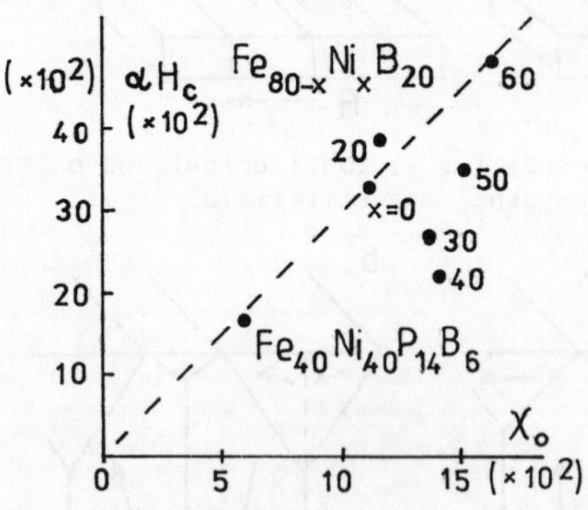

Fig.7. Correlation between αH_c and χ_o for various FeNi-based amorphous alloys [9].

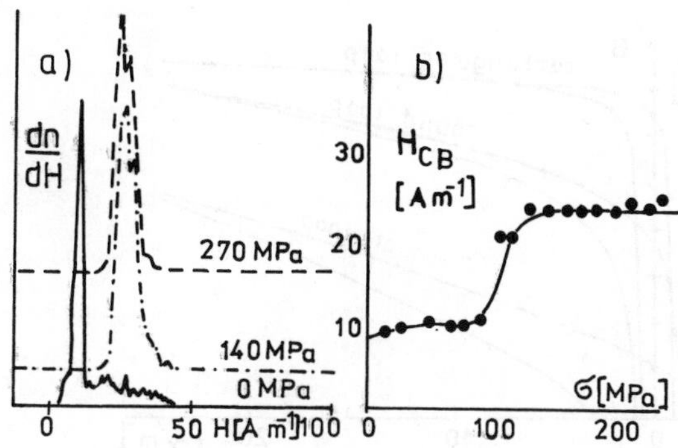

Fig.8. a) Field dependences of number of registrated Barkhausen jumps dn/dH for various values of tensile stress and b) stress dependence of Barkhausen coercive field H_{cB} in $Fe_{60}Ni_{20}B_{20}$ amorphous alloy [13].

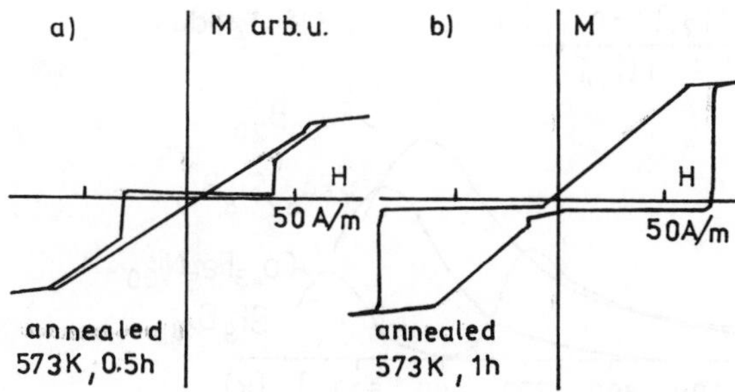

Fig.9. Hysteresis loops of $(Fe_{.06}Co_{.94})_{79}Si_2B_{19}$ amorphous ribbons annealed for 30 min (a) and 1 hour (b) in a demagnetized state[14].

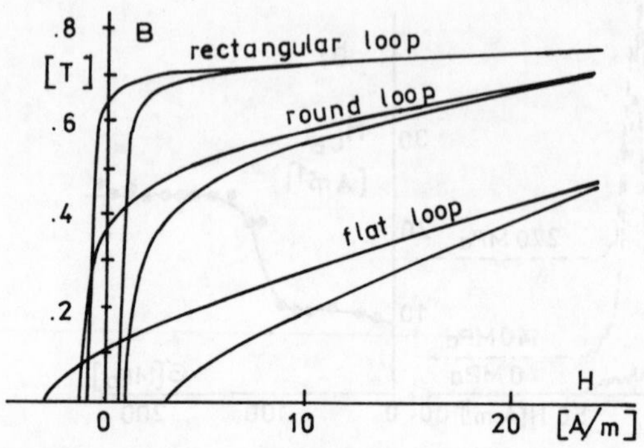

Fig.10. Statical hysteresis loops of strip-wound cores of $Fe_{39}Ni_{39}Mo_4Si_6B_{12}$ alloy after various annealing treatments[16].

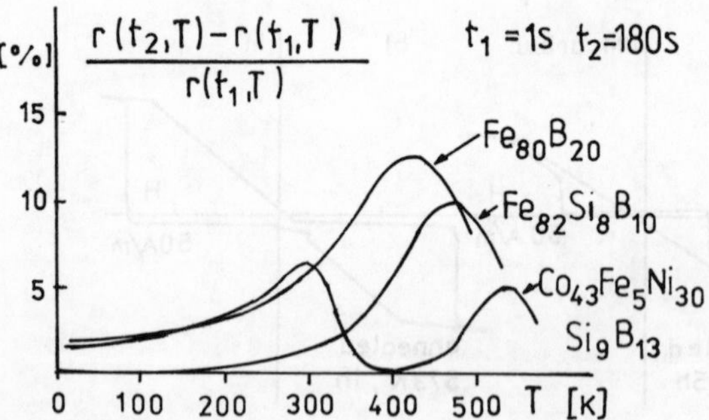

Fig.11. Relaxation spectra of three characteristic amorphous alloys showing the temperature dependence of magnetic aftereffect[22].

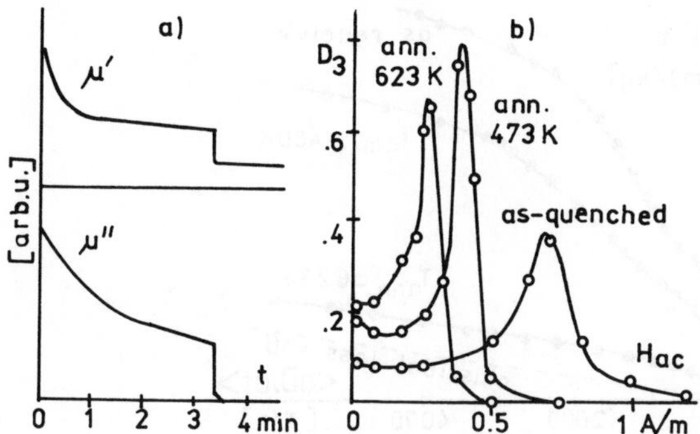

Fig.12. a) Anomalous drop of both components of complex permeability in $(Fe_{.05}Co_{.95})_{78}Si_8B_{14}$ alloy [24]. b) Influence of annealing on field dependence of disaccommodation D_3 (measured 3 min after demagnetization) in $Fe_{4.7}Co_{70.3}Si_{15}B_{10}$ amorphous alloy [25].

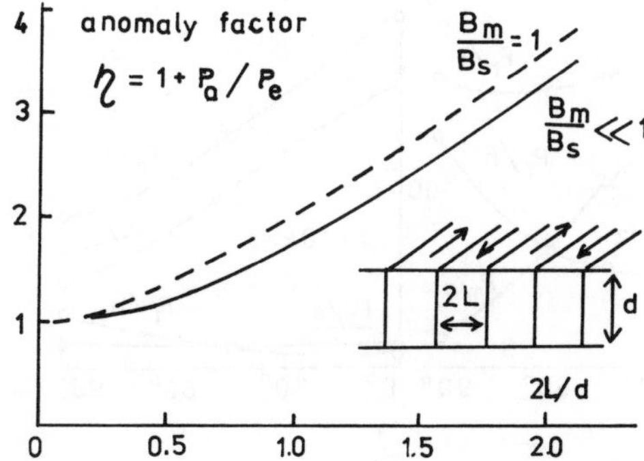

Fig.13. Anomaly factor η as a function of the ratio of the domain wall spacing 2L to the sheet thickness d, calculated by Pry and Bean [28].

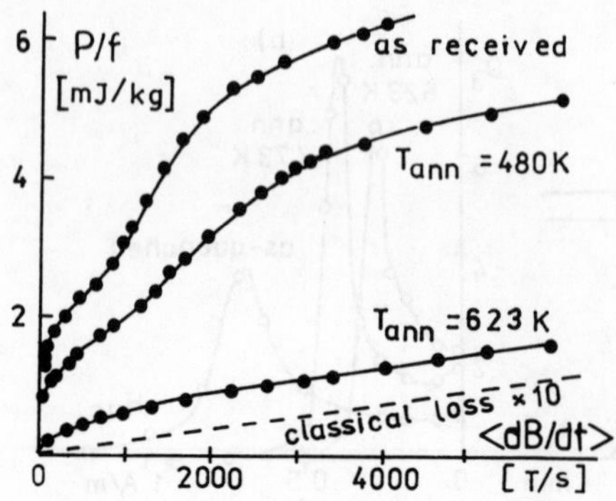

Fig.14. The influence of annealing on frequency dependence of power losses per cycle of $Fe_{40}Ni_{40}P_{14}B_6$ amorphous alloy [30]. Classical loss is shown (dashed line) in an expanded scale.

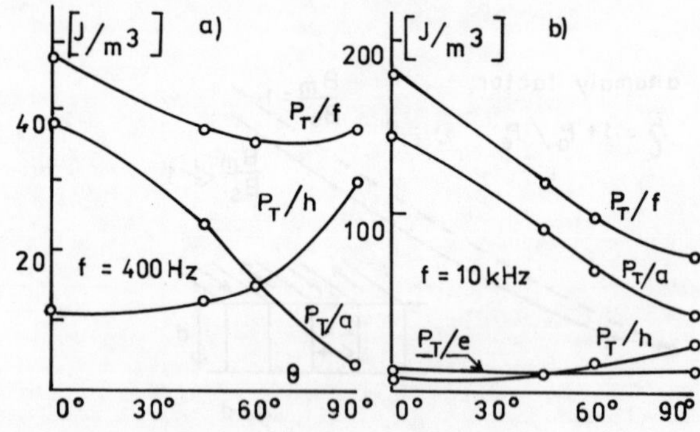

Fig.15. Total loss per cycle P_T/f and constituent parts: p_h/f, P_e/f and P_a/f as a function of angle θ between the direction of annealing field and the ribbon length. Measurements are performed on $Fe_{78}Si_{10}B_{12}$ ribbons at B_m = 1T and frequencies: a) - 400 Hz and b) - 10 kHz [32].

RAPIDLY SOLIDIFIED HARD MAGNETIC MATERIALS

R. Grössinger, G. Hilscher

Institut für Experimentalphysik, Technical University, Vienna, Austria

1. INTRODUCTION

Basic compounds which gain interest as high performance hard magnetic materials should be ferromagnetically ordered with a high Curie temperature and a large saturation magnetization, additionally they should exhibit a high uniaxial magnetic anisotropy at room temperature. The occurence of a high magnetocrystalline anisotropy is supported by a crystalline structure with a uniaxial symmetry (as e.g. the hexagonal, rhomboedric or tetragonal structure). Furthermore magnetic ions with a non-vanishing orbital moment (e.g. rare earths) should be a part of the compound composition. Famous intermetallics are here $SmCo_5$ and Sm_2Co_{17} which at room temperature exhibit anisotropy fields of 350 kOe and 70 kOe, respectively. According to the fact that the magnetocrystalline anisotropy is mainly caused by the crystal field acting upon the orbital moment of the 4f-electrons, the symmetry of the local surrounding is important for the magnitude as well as the sign of the anisotropy. Therefore in the crystalline state the magnetic anisotropy can successfully be calculated using a point charge model [1,2].

In the amorphous state the situation is much more complex. If a material is rapidly solidified with cooling rates above 10^6 K/s, generally the amorphous state (which can be understood as an undercooled liquid with short range order effects) is formed. The large number of soft magnetic 3d-metalloid compounds (e.g. $Fe_{80}B_{20}$, $Co_{80}B_{20}$) are well-known in this respect, where the whole magnetization process is determined by the interaction of the magnetostriction with local stresses (see e.g. [3]). Amorphous samples containing rare earth exhibit anisotropy effects (difficulties to saturate the material, broad hysteresis) due to the randomly oriented local crystal field which defines the local easy exis at each rare earth site. However, these compounds generally show low ordering temperatures indicating a reduced exchange. Therefore these compounds are generally not of interest for technical applications.

Necessary conditions for a compound, which may be applicable for a technical permanent magnet are:

i) A high saturation magnetization.

ii) An ordering temperature far above room temperature.

iii) A high uniaxial magnetic anisotropy.

All these conditions were up to now only fulfilled by crystalline compounds. However, it was shown recently, that the quenching procedure fast cooling of melt can be used to produce materials which show a microcrystalline structure having excellent hard magnetic properties even at room temperature [4]. The composition of these magnets were close to $Nd_{15}Fe_{77}B_8$, however, the stochiometry of the grains responsible for this behaviour is $Nd_2Fe_{14}B$. This compound is of the tetragonal structure (space group $P4_2/mnm$) [5], and exhibits a high saturation magnetization combined with a uniaxial anisotropy at room temperature [6]. The most important magnetic data of the rare earth-3d compounds which attracted growing interest as high performance permanent magnets are summarised in Table 1.

Sm-Co based magnets have excellent hard magnetic properties (energy products of 30MGOe were achieved!), however the high costs of the raw materials are prohibitive for large scale applications. R-Fe-B based magnets are much cheaper, additionally R=Nd,Pr or Mischmetal are much more abundant than Sm. The only disadvantage is the rather low Curie temperature of $Nd_2Fe_{14}B$, which deteriorates the thermal stability.

2. MAGNETIC PROPERTIES OF R-Fe-B COMPOUNDS

$R_2Fe_{14}B$ compounds are formed with all rare earth elements. At room temperature the materials with R = Y, La, Ce, Pr, Nd, Gd, Tb, Dy, Ho, Lu have an easy c-axis of magnetization [7]. For technical applications both sintered and rapidly quenched Nd-Fe-B magnets were produced [4,8]. From the study of the R-Fe-B compounds with R = Y, La, Ce, Lu the anisotropy behaviour of the 3d-sublattice can be deduced. Fig. 1 shows the temperature dependence of the anisotropy field between 4.2 K and the Curie temperature of nearly all $R_2Fe_{14}B$ according to [9]. $Nd_2Fe_{14}B$ is uniaxial at room temperature, but below 135 K a spin-reorientation to an easy cone occurs [10]. Between 180 K and 135 K a step in the M(H) curve was detected, indicating a FOMP (First Order Magnetization Process) transition [9]. The physical origin of this anomaly is, that K_1 (first order anisotropy constant) changes at the spin-reorientation its sign. Consequently, K_1 decreases if this critical temperature at 135 K is approached. The influence of the higher order anisotropy is therefore no longer negligible. The higher order anisotropy constants can cause additional relative minima in the anisotropy energy which can lead to such jumps in the M(H) curve. For the hexagonal symmetry the corresponding theory was developped by [11], for the tetragonal case an analysis was given by [12]. Such a FOMP was detected for the first time in $PrCo_5$, which shows also at lower temperatures a spin-reorientation [13]. The fact that this behaviour was only detected in Nd containing R-Fe-B compounds points to the fact that the Nd-sublattice is responsible for this anisotropy curiosity. It is worth to note that Nd_2Fe_{17} has a Curie temperature of 339 K whereas $Nd_2Fe_{14}B$ orders at 595 K. This means that the Fe-Fe exchange is drastically enhanced in this structure. Summarising the role of the elements involved: Fe is due to its large exchange interaction responsible for the high T_c value. B is necessary to stabilise this tetragonal structure. The rare earth elements are necessary for a high uniaxial anisotropy. Especially the light rare earths are important because only their moments couple ferromagnetically with Fe, whereas an antiparallel 4f-3d coupling occurs for the heavy rare earths.

Studies performed on Nd-Fe-B samples with various stoichiometries showed that this structure has a small homogeneity range and appears to be very stable. The magnetic anisotropy, which is very sensitive to the local surrounding, was insensitive to any variation of the stochiometry[12], but also to any heat treatment of the samples [14]. This is not trivial, because in Sm-Co based magnets the anisotropy depends dramatically on such variations [15].

3. R-Fe-B BASED MAGNETS

The general composition used for technical permanent magnets is $Nd_{15}Fe_{77}B_8$. These magnets can be produced in two different ways:

i) The polycrystalline material is powdered to a mean grain size of 5-10 μm, pressed under an external field in a die, sintered at 1100^oC and heat treated at approximately 600^oC. The grain size is here larger as the critical single domain diameter [16]. This technique is very similar to that generally used for $SmCo_5$ magnets. Anisotropic magnets with energy products above 40 MGOe were achieved.

ii) If the same polycrystalline Nd-Fe-B material is rapidly quenched using a single roller technique, pieces of brittle ribbons are obtained, which are microcrystalline with a grain size much smaller than the critical diameter of a single domain particle [16]. The thus obtained material is naturally isotropic. The cooling rate can continously be changed by varying the speed of the copper wheel which is used for the single roller technique. Consequently, this technique allows to vary the grain size, which is of great importance for the coercivity. Fig. 2 demonstrates this by plotting ${}_IH_c$ as measured at room temperature as a function of the wheel velocity for two different compositions ($Nd_{15}Fe_{77}B_8$ and $Nd_{14}Fe_{81}B_5$) [17]. ${}_IH_c$ was measured on plastic bonded isotropic magnets statically as well as on the ribbons directly in a pulsed field. It should be mentioned that the anisotropy field as obtained on the ribbons did not depend on the wheel velocity and is found to remain almost constant at about 75 kG.

The fact that with both techniques in principle the same compound is formed is demonstrated by an X-ray diffraction pattern which was obtained on polycrystalline $Nd_{15}Fe_{77}B_8$ but also on rapidly quenched material with two different cooling rates (see Fig. 3). The characteristic lines of the $Nd_2Fe_{14}B$ structure are always visible, the rapid quenching causes only a line broadening. It is therefore clear that the hard magnetic properties are in both cases caused by the $Nd_2Fe_{14}B$ grains.

4. COMPARISON OF THE MAGNETIC PROPERTIES OF SINTERED AND RAPIDLY QUENCHED Nd-Fe-B MAGNETS

Beside a high saturation magnetization, responsible for a high remanence, a high coercivity field is desired for high quality permanent magnets. In the case of an ideal single domain particle magnet the anisotropy field represents the theoretical upper limit for the coercive field. This theoretical value corresponds to a coherent rotation of the magnetization in perfect single domain particles. For a real technical permanent magnet, however, ${}_IH_c$ values of approximately 10% of H_A are achieved. The question now arises why ${}_IH_c$ is much lower than its theoretical limit. According to [18,19] two different mechanisms are responsible for magnetic hardening:

i) Domains of reversed magnetization are difficult to nucleate by an external field (i.e. the nucleation field H_n), while existing domain walls can easily be moved.

ii) The mobility of domain walls is hindered by various defects which act as pinning centres, although many nuclei with reversed magnetization exist in the crystal.

By measuring the virgin curve as well as successive minor loops (this means hysteresis loops with increasing external fields) starting from the thermal demagnetized state, it is possible to distinguish between these two models. In the first case where the magnetization reversal is controlled by nucleation the coercivity is determined by the nucleation field H_n which is reduced by the local stray fields at the grain boundary. According to Herzer et al. [20] ${}_IH_c$ can be written:

$$_IH_c = cH_n - aM_s$$

where c accounts for the exchange interaction across the grain boundary (for isolated grains $c \to 1$ while for strongly coupled grains $c \to 0$) and "a" represents the "demagnetizing factor", which may attain $8\pi M_s$ for reversed grains facing each other[20]. In this material the virgin curve increases rapidly due to the easy movement of the domain walls and the saturation is attained already at small external fields, while the coercivity raises almost linearly with the magnetizing peak field. It approaches the maximum coercive field according to the above relation. A typical example for a nucleation dominated magnet are the minor loops as measured on a $Nd_{15}Fe_{77}B_8$ sintered magnet (see Fig. 4a). In the second type of materials which are controlled by a pinning mechanism the magnetization of the virgin curve remains very small up to the pinning field H_p and increases afterwards steeply to saturation. The coercivity is almost independent of the external field for $H_{ext} > H_p$. If Nd is partly replaced by heavy rare earth elements as Tb or Dy in order to enhance the anisotropy, we observe for H_{ext} larger than 7-10 kG successive minor loops which are typical for a pinning mechanism (see Fig.4b [21]). At low fields, however, the firstly steep rising minor loops after thermal demagnetization indicate the immediate growth of domains due to an easy movement of the domain walls within the grain. With increasing external field the slope of the successive minor loops flattens which indicates that pinning centres as grain boundaries become active and thus reduce the mobility of the domain walls. In this respect the behaviour of rapidly quenched magnet is interesting. Fig.5a shows minor loops of an over-quenched sample with a wheel velocity v > 20 m/s (see right hand side of Fig.2). The hysteresis loop of this sample is typical for a multi-phase material. After heat treatment a high coercivity develops and the shape of the minor loops is typical for a pinning controlled magnet. It should be pointed out that the composition of this sample was $Nd_{15}Fe_{77}B_8$, the same as generally used for sintered materials. The different production procedure has changed the coercivity mechanism drastically. This demonstrates the importance of different grain sizes but also of different additional phases for the coercivity.

It is interesting to note that the temperature dependence of the anisotropy behaves similar for both the rapidly quenched material and the sintered magnets. Fig.6 shows the temperature dependence of the anisotropy field as well as that of the coercivity of rapidly quenched $Nd_{15}Fe_{77}B_8$. Below 200 K the typical anisotropy anomaly (FOMP) characteristic for this compound could be detected. It should be noted that at this temperature a downturn of the coercivity field was observed. This is due to the fact that rapidly quenched material is isotropic. Regarding the temperature dependence of $_IH_c$ of sintered magnets two cases exist:

i) Applying the external field parallel to the preferential axis, a smoothly increasing $_IH_c$ with decreasing temperature was detected.

ii) Applying the external field perpendicular to the aligning axis, a drastic downturn below 200 K of $_IH_c(T)$ was observed [22].

It is obvious that in the case of an isotropic material this two effects are superimposed. This explains, in principle, the behaviour of the $_IH_c(T)$ curve of the rapidly quenched material.

Another possibility to distinguish the various coercivity mechanisms is based on an analysis of the temperature dependence of the anisotropy field and the coercivity field. The basic idea is that the coercivity can be correlated with the magnetocrystalline anisotropy. Kütterer et al. [23] have shown that in a first approximation, neglecting K_2, H_n is given by $2K_1/M_s$ which, again neglecting K_2, is also the expression for the anisotropy field H_A. Since the nucleation field cannot be measured directly we used for our analysis the real anisotropy field H_A instead of H_n and assumed the general relation of the following type: $_IH_c(T)$ prop. $(H_A(T))^k$. The value of the power "k" is an important parameter which is determined by the coercivity mechanism. According to Kütterer et al. [23] $k = 3/2$ appears to be characteristic for a pinning of domain walls at grain boundaries if the width of the domain walls is larger than the grain boundaries. $k = 5/2$ is characteristic for volume pinning which means the pinning centres are randomly distributed within the grain. $k = 5/4$ is indicative for a pinning of narrow domain walls at atomic defects. A plot of $\ln({}_IH_c(T)/{}_IH_c(T = 300\ K))$ divided by $\ln(H_A(T)/H_A(T = 300\ K))$ vs the temperature shows that "k" changes from approximately 3/2 at low temperatures to 5/2 around room temperature in sintered Nd-Fe-B magnets (see Fig.7 according to [24]). This can be explained by a pinning of domain walls on grain boundaries, where the dimension of the domain walls is broad compared with that of the grain boundaries. At higher temperatures the thickness of the domain walls increases, leading to a nucleation of these walls at inclusions in the grains. It should be mentioned that the mean grain size is large compared to the critical single domain diameter [16]. However, the change of "k" may also be a consequence of the rising importance of K_2 as the temperature is lowered since K_1 changes its sign at about 135 K. According to a recent calculation of Herzer et al. [20] the nucleation field H_n depends strongly on K_2 for $-2K_2<K_1<4K_2$. In this range H_n is always smaller than H_A.

A similar analysis of rapidly quenched Nd-Fe-B material gives a "k" value close to 1. A "k" of 5/4 indicates, according to [23], a pinning of the domain walls on statistically distributed atomic defects. In the case of rapidly quenched material the mean grain size is much smaller than the critical single domain particle diameter and atomic defects within the grains become active as pinning centres. The grain boundaries in sintered magnets compared with the atomic defects in rapidly quenched magnets lead to a well distinguishable coercivity mechanism.

5. CONSLUSION

In sintered Nd-Fe-B magnets minor loop measurements performed at room temperature showed a nucleation dominated coercivity mechanism. Rapidly quenched material behaves differently, the minor loops point to a pinning of domain walls, which might be responsible for the coercivity. These assumptions are in principle in agreement with an analysis of the temperature dependence of the anisotropy field, which was correlated with that of the coercivity. In sintered magnets at low temperatures a pinning of broad domain walls at grain boundaries and/or the rising importance of K_2 was concluded. At higher temperatures a nucleation mechanism of the domain walls on inclusions plays the most important role. On the contrary in rapidly quenched Nd-Fe-B a different coercivity mechanism was found, which is based on the pinning of the domain walls on statistically distributed small defects. This means that both experimental techniques (minor loops; measuremen of ${}_{I}H_{c}(T)$ and $H_A(T)$) lead in principle to the same theoretical picture. The most interesting conclusion of these studies is, that the anisotropy of the $Nd_2Fe_{14}B$ phase is the origin of the coercivity in sintered and rapidly quenched Nd-Fe-B magnets. The metallurgy of the production techniques affect via different grain sizes and additional phases the coercivity mechanism. The magnetic properties of these impurity phases seem to be of minor importance.

REFERENCES

1) Hutchings, M.T., Sol. State Phys., 16, 227 (1966).
2) Callen, E., Physica, 114B, 71 (1982).
3) Grössinger, Sassik, H., Schotzko, Ch., Veider, A., Z. f. Metallkunde, 74, 577 (1983).
4) Croat, J.J., Herbst, J.F., Lee, R.W., Pinkerton, F.E., J. Applied Phys., 55, 2078 (1984).
5) Herbst, J.F., Croat, J.J., Pinkeron, F.E., Yelon, W.P., Phys. Rev., B29, 4176 (1984).
6) Grössinger, R., Obitsch, P., Sun, X.K., Eibler, R., Kirchmayr, H.R., Rothwarf, F., Sassik, H., Mat. Lett., 2, 539 (1984).
7) Sinnema, S., Radwanski, R.H., Franse, J.J.M., De Mooij, D.B., Buschow, K.H.J., J. Magn. Magn. Mat., 44, 333 (1984).
8) Sagawa, M., Fujimura, S., Togawa, M., Yamamoto, H., Matsuura, Y., J. Applied Phys., 55, 2083 (1984)
9) Grössinger, R., Sun, X.K., Eibler, R., Buschow, K.H.J., Kirchmayr, H.R., J. de Physique C6(9), C6-221 (1985).
10) Givord, P., Li, H.S., Moreau, J.M., R. Perrier de la Bathie E. du Tremolet de Lacheisserie, Physica 130B, 323 (1985).
11) Asti, G., Bolzoni, F. J. Magn. Magn. Mat., 15-18, 29 (1980).

12) Sun, X.K., Thesis, Techn. Univ., Vienna, (1985).

13) Asti, G., Bolzoni, F., Leccabue, F., Panizzierie, R., Pereti, L., Rinaldi, S., J. Magn. Magn. Mat., 15-18, 561 (1980).

14) Grössinger, R., Kirchmayr, H.R., Krewenka, R., Narasimhan, K.S.V. L., Sagawa, M., Proc. of 8-th Int. Workshop on rare earth magnets Dayton (USA) p.553 (1985).

15) Grössinger, R., Obitsch, P., Kirchmayr, H., Rothwarf, F., IEEE Trans. on Magn. MAG-20, 1575 (1984).

16) Durst, K.D., Kronmüller, H., J. Magn. Magn. Mat. (1986), in print.

17) Hilscher, G., Grössinger, R., Heisz, S., Sassik, H., Wiesinger, G., J. Magn. Magn. Mat., 54-57, 577 (1986).

18) Becker, J.J., IEEE Trans. on Magn. MAG-18, 1451 (1982).

19) Menth, A., Nagel, A., Perkins, R.S., Ann. Rev. Mater. Sci., 8, 21 (1978).

20) Herzer, G., Fernengel, W., Adler, E., J. Magn. Magn. Mat., 58, 48 (1986).

21) Heisz, S., Diplomarbeit T.U., Wien (1985).

22) Grossinger, R., Krewenka, R., Narasimhan, K.S.V.L., Segawa, M., J. Magn. Magn. Mat., 51, 160 (1985).

23) Kütterer, R., Hilzinger, H.R., Kronmüller, H., J. Magn. Magn. Mat., 4, 1 (1977).

24) Grössinger, R., Krewenka, R., Narasimhan, K.S.V.L., Kirchmayr, H. R., Proc. of Intermag Conf., Phoenix, USA (1986).

Table 1:

Saturation magnetization, anisotropy fields (at room temperature) and Curie temperatures of for permanent magnets important material.

Compound	$4\pi M_s$ (kG)	H_A (kG)	T_c (K)
$SmCo_5$	11.2	290-350	1020
Sm_2Co_{17}	12.8	65-90	1195
$Nd_2Fe_{14}B$	16	75	580

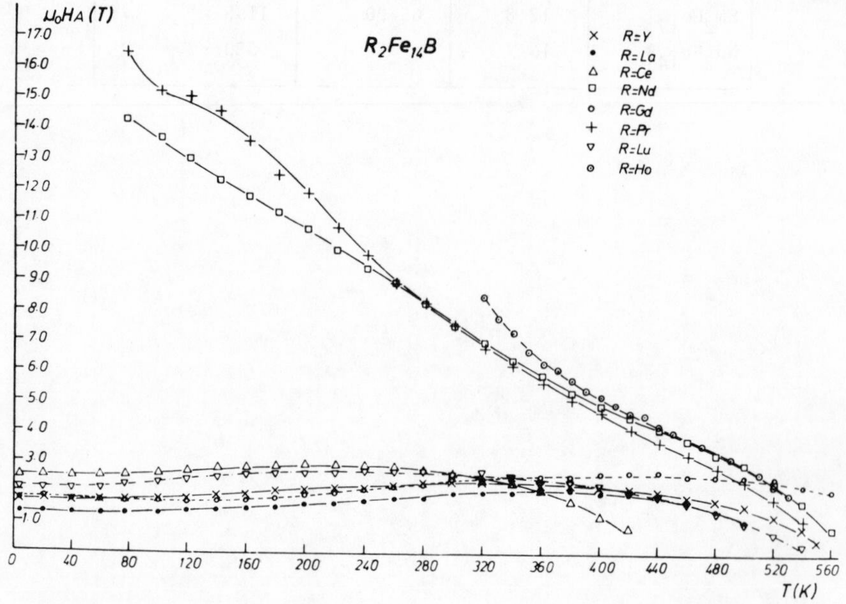

Fig.1. Temperature dependence of the anisotropy field $\mu_o H_A$ of the $R_2Fe_{14}B$ (R=Y,La,Ce,Nd,Pr,Gd,Ho,Lu) compounds.

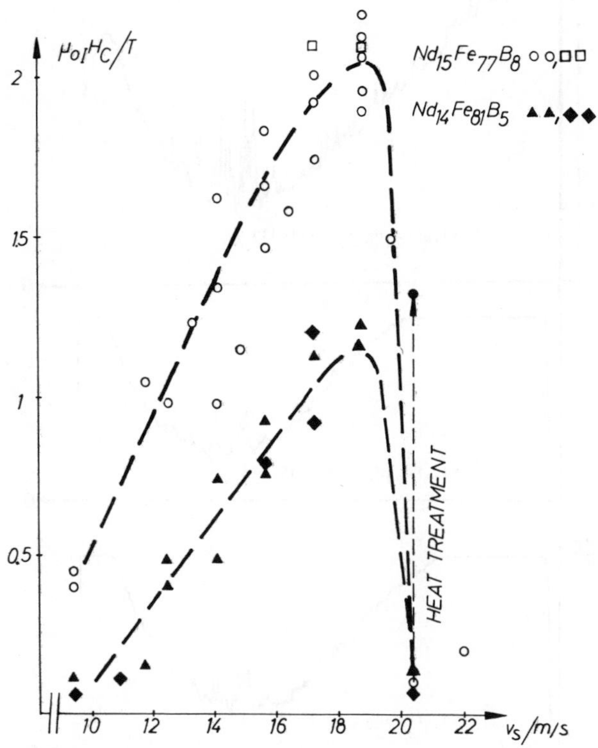

Fig.2. The dependence of $_IH_c$ of melt-spun ribbons upon the wheel velocity. Static measurements $Nd_{15}Fe_{77}B_8$ (o), $Nd_{14}Fe_{81}B_5$ (▲), pulsed field measurements (◘ ,◆).

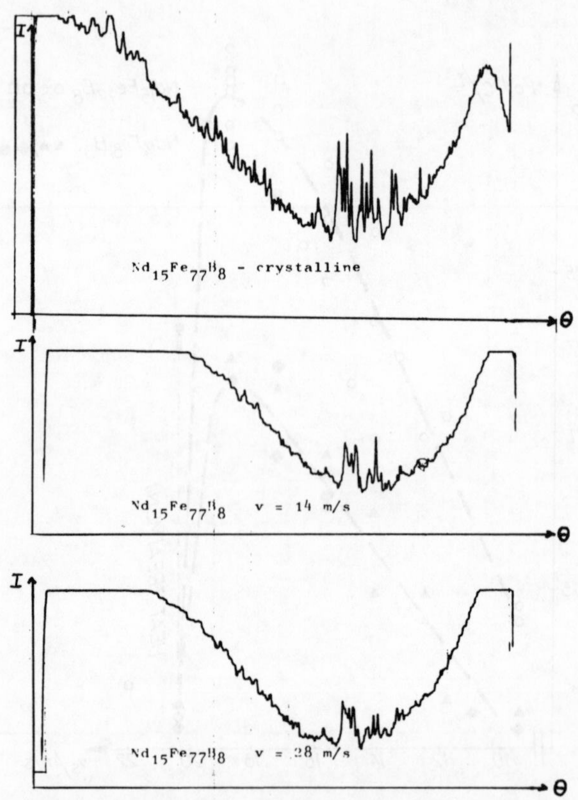

Fig.3. X-ray diffraction pattern on a) polycrystalline $Nd_{15}Fe_{77}B_8$ and on rapidly quenched $Nd_{15}Fe_{77}B_8$ b) wheel velocity v = 14 m/s, c) wheel velocity v = 28 m/s obtained with Cu-Kα radiation.

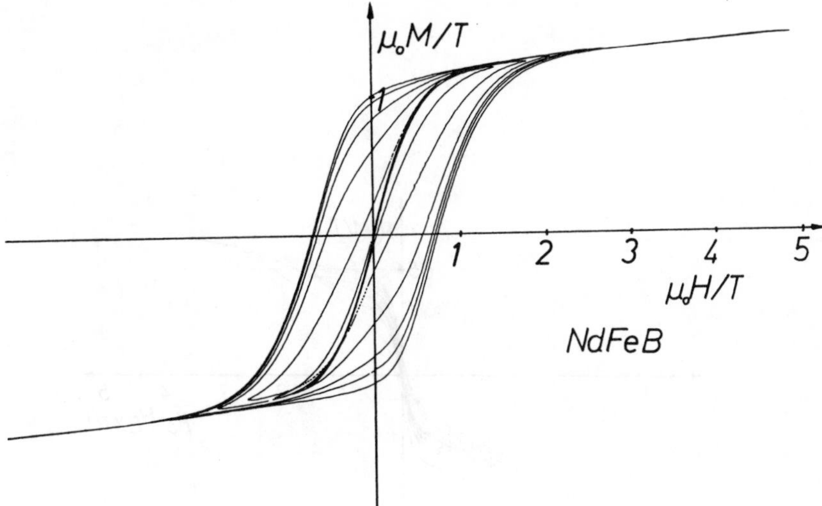

Fig.4a: Minor loops as obtained on a commercial sintered Nd-Fe-B magnet.

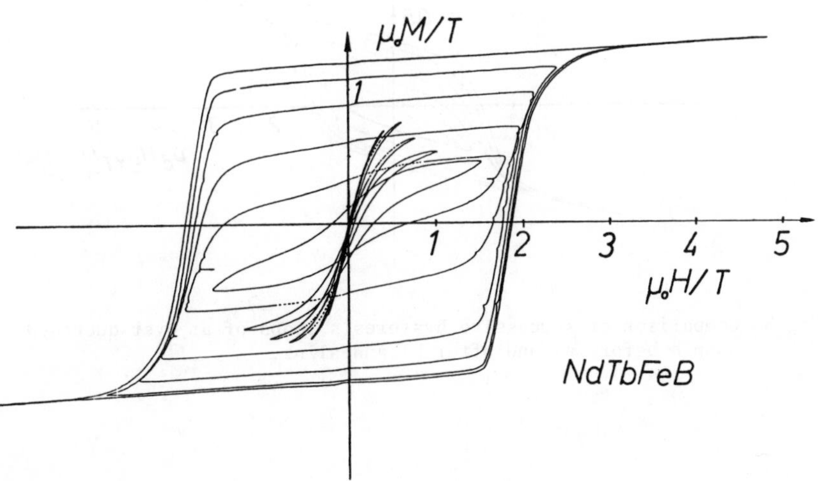

Fig.4b: Minor loops as obtained on a commercial sintered Nd-Fe-B magnet, where 10% of the Nd was substituted by Tb.

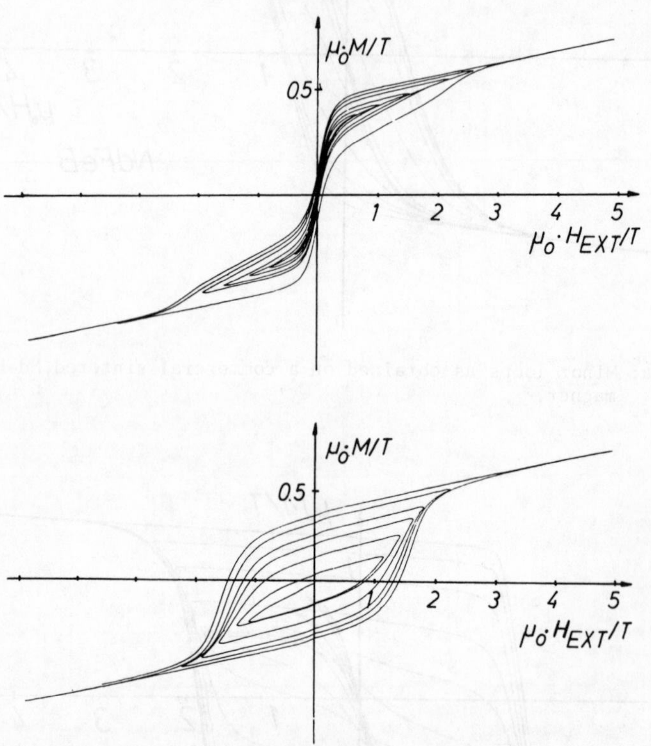

Fig.5. Comparison of successive hysteresis loops of an over-quenched sample before a) and after b) annealing.

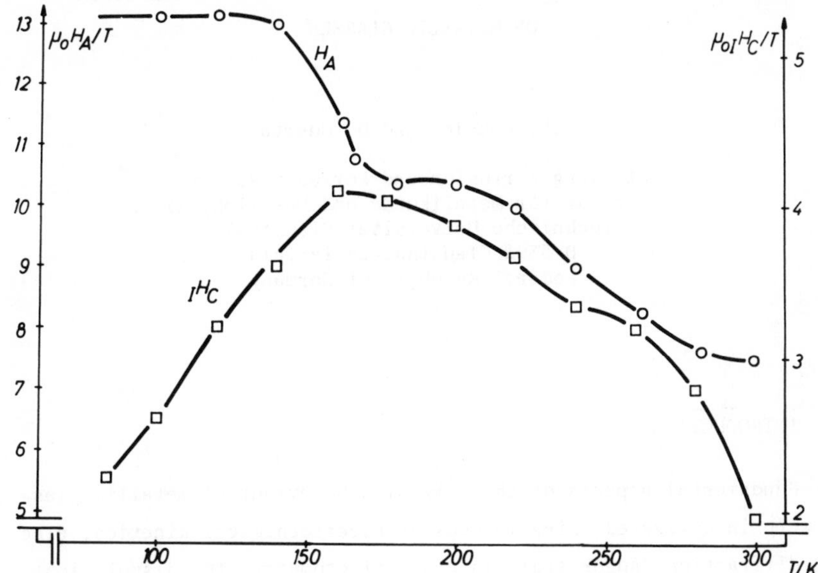

Fig.6. Temperature dependence of the anisotropy field (o) and coercivity (□) of a melt spun ribbon v = 15 m/s.

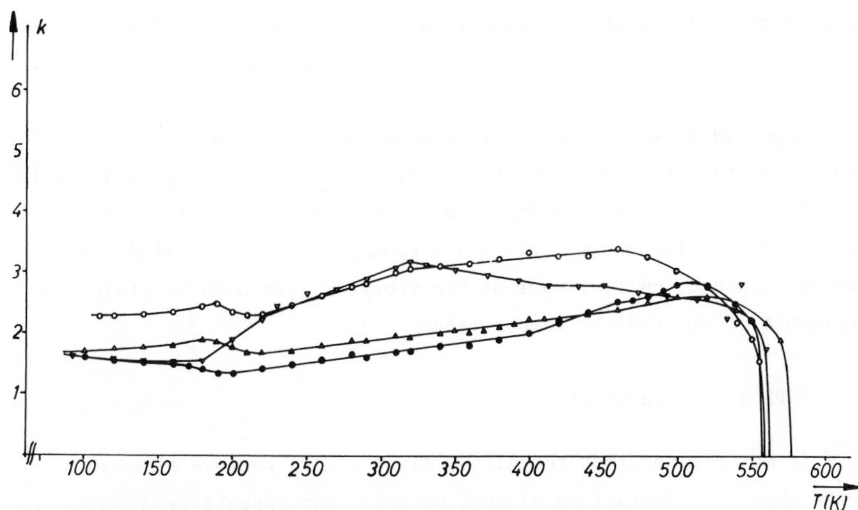

Fig.7. Temperature dependence of the power factor $k = (\ln {_I}H_c(T)/{_I}H_c(T{=}300K))/\ln(H_A(T)/H_A(T{=}300K))$ for various Nd-Fe-B magnets (∇,o,Δ and a heavy rare earth substituted magnet (●).

INFLUENCE OF COMPOSITION AND STRUCTURE ON THE CORROSION OF METALLIC GLASSES

K.E. Heusler and D. Huerta

Abteilung Korrosion und Korrosionsschutz
Institut fur Metallkunde und Metallphysik
Technische Universitat Clausthal
D 3392 Clausthal-Zellerfeld
Federal Republic of Germany

1. INTRODUCTION

Fundamental aspects of the corrosion behaviour of metallic glasses were investigated using methods of electrochemical kinetics, electron diffraction, Auger analysis and reflectometry. The dissolution kinetics of glasses was compared to that of homogeneous crystals of the same composition. Selective dissolution of components and the influence of substitution of components in the glass on the dissolution kinetics were studied. Some part of this work was published previously[1-3].

Experiments were mainly performed with binary and ternary boron glasses containing iron, cobalt, nickel, vanadium, chromium and platinum. PdSi, PdZr, NiZr, and FeZr glasses were also used. The corrosion behaviour of different materials was compared using acidified 0.5 M sodium sulphate, pH 1.8. Various electrolytes were used to study the influence of electrolyte composition.

2. ASPECTS OF STRUCTURE

It was claimed that the glassy metal alloys corrode at slower rates than the crystalline alloys. However, one expects from theoretical considerations that anodic dissolution rates of glassy alloys corroding in the active state should be higher than those of homogeneous crystals of the same composition. Since the homogeneous crystal is closer to the equilibrium state than the glass, activity of the components will be smaller in the crystal.

On the other hand, the corrosion rate of a heterogeneous crystalline alloy can be higher than that of a glass of the same mean composition. The reason will be the contact corrosion of the phase with free corrosion potential more negative than that of another phase.

No general predictions can be made for the effect of structure on the rate of the cathodic reduction of some oxidizing substance in the electrolyte, but any effect should be small. Similarly, one will not expect much changes of the corrosion behaviour in the passive state of an alloy due to its structure.

Upon heat treatment $Ni_{66}B_{34}$ crystallizes forming the thermodynamically stable homogeneous phase. The dissolution rate of the crystalline alloy at a given electrode potential was observed to be of about one order of magnitude smaller than the dissolution rate of the glass.

In other systems the dissolution rate decreased at first, due to the heat treatment and then increased after further heat treatment. This effect was observed for $Fe_{75}B_{25}$ and for transpassive $Ni_{25}Zr_{75}$, but most clearly for $Co_{75}B_{25}$.

Fig.1 shows the dissolution rates at -0.3 V vs. SCE after 5 h heat treatment at the indicated temperatures. The dissolution rates dropped sharply in the temperature range between 570 K and 610 K. At each temperature electron diffraction patterns were taken. At temperatures up to 550 K diffraction patterns typical of a glass were observed. The first indications of crystallization appeared at 570 K. At 610 K well developed diffraction patterns of the tetragonal Co_3B-phase were found by the selected area diffraction. After heat treatment at higher temperatures the diffraction patterns of additional phases were observed indicating decomposition of Co_3B into Co_2B and Co. Analogous results were obtained when the heat treatment was performed at 670 K for different times. The experiments clearly demonstrate the validity of theoretical expectations.

For the crystalline metals: iron, cobalt or nickel it was found that anomalous Tafel slopes of about 30 mV/decade corresponding to

RT/(1+2β)F with transfer coefficients β close to 0.5, and reaction orders in excess of one with respect to hydroxyl ions in the electrolyte, were due to the almost exclusive dissolution of metal ions from kinks in monoatomic steps and to the potential dependent and the pH dependent steady state surface concentration of kinks [4,5]. Since there are no steps and kinks on the surface of glass, but only a continuous spectrum of differently bound atoms, the Tafel slope should be RT/2βF or about 60 mV/decade, and the reaction order with respect to hydroxyl ions should be equal one. This is exactly the result obtained from experiments with $Fe_{80}B_{20}$ shown in Figs. 2 and 3. After crystallization the Tafel slopes became smaller, a tendency which was also observed, e.g. for CoB glasses.

3. ASPECTS OF ALLOY COMPOSITION

Electrochemical rate constants for the dissolution of components of an alloy may differ by many orders of magnitude. As a consequence, the composition of the alloy at the surface will change with time. If practically only one component is dissolved and if diffusion in the alloy is sufficiently fast, changes of composition penetrate into macroscopic distances from the surface. Then one observes a current limited by diffusion in the alloy. From the diffusion limited current one may determine the diffusion coefficients.
Boron from a FeNiB glass dissolves only at negative electrode potentials at which dissolution of the other components is negligibly slow. Diffusion coefficients around 10^{-16} cm^2/s were calculated from the diffusion limited current, but also from the profile of the composition at the surface obtained by Auger analysis, and from reflectometry with polarized monochromatic light [1].

In Fig.4 the reciprocal current densities for dissolution of boron from a PtB glass are plotted vs. square root of time, according to the relation

$$j(t)^{1/2} = nF(x_B/V)(D/\pi)^{1/2}. \qquad (1)$$

In eq. (1) j is the current density, t - time, n - the charge number which assumes value $n = 3$ for boron, x_B - the molar fraction of boron, V - the molar volume of the alloy, and D - the chemical diffusion coefficient. The experiments in Fig. 4 yield $D = 2\ (\pm 1) \cdot 10^{-13}\ cm^2/s$.

Glassy alloys offer an opportunity to change the composition of a phase in wide range, while the ranges of existence of homogeneous crystalline phases are often severely limited. Thus, it is possible to test a simple theory on dissolution of alloys in which rate constants for the dissolution of the components are assumed to be independent of composition. For a binary alloy the total current density j can be written as the sum of the component current densities j_i:

$$j = \sum_i j_i . \tag{2}$$

Each component of current density is given by the concentration of a component at the surface expressed by the molar fraction x_i^s and the rate constant k_i:

$$j_i = k_i x_i^s . \tag{3}$$

In steady state an alloy dissolves completely, if the ratio of the component fluxes is equal to the ratio of the bulk molar fraction. Since the flux is given by j_i/n_i, one obtains for a binary alloy with the components A and B, setting $x_A = x$

$$j_A/j_B = n_A x/n_B(1-x) . \tag{4}$$

The charge numbers n_i are those for the reactions:

$$M_i = M_i^{(n_i)+} + n_i e^- . \tag{5}$$

Elimination of the usually unknown x^s yields:

$$j = (n_A x + n_B(1-x))/((n_A x/k_A) + n_B(1-x)/k_B) . \tag{6}$$

The general equation for multicomponent systems is

$$j = (\sum_i n_i x_i)/(\sum_i n_i x_i/k_i) . \tag{7}$$

In an ideal system the rate constants k_i are independent of the compositions at the surface and in the bulk. The rate constant for the pure components can be measured, at least in principle. Thus, for an ideal system eqs. (6) and (7) predict the current densities as a function of composition.

For binary glasses Fe_yB_{1-y} one may rewrite eq. (6) with $n_{Fe} = 2$ and $n_B = 3$ as:

$$(3-y)/j = y((2/k_{Fe})-(3/k_B))+(3/k_B) \quad . \tag{8}$$

Fig.5 shows experimental results for $0.75 \leqslant y \leqslant 0.84$. The linear relation between $(3-y)/j$ and y indicates that the assumption of an ideal behaviour is justified in the relatively narrow range of concentrations. From the intercept at y = 0 one finds $k_B = 0.78\ mA/cm^2$ and from the slope $k_{Fe} = 9.5\ mA/cm^2$ at E = -0.5 V vs SCE.

In ternary glasses of the type $(A_{1-z}M_z)_{1-y}B_y$ one metal A may be substituted by another metal M. The degree of substitution is defined by z. Using eq. (7), the ratio of the current density j(z=0) to the current density j(z), at different degrees of substitution, j(z) for some constant boron concentration y, $n_A = n_M = 2$ and $n_B = 3$ is given by:

$$\left(\frac{j(z=0)}{j(z)}\right)_{y=const} = \frac{1+(1-y)z((k_A/k_M)-1)}{1.5y(k_A/k_B)+1-y} \tag{9}$$

Experiments were performed with iron-boron glasses for y = 0.13 in which iron was substituted by Co, Ni, V, Mo or Mn. Then eq.(9) reduces to:

$$j(z=0)/j(z) = 1 + 0.257z((k_{Fe}/k_M) - 1 \tag{10}$$

The experiments showed that the steady state dissolution of the active glassy alloys was described by Tafel lines with the slope being independent of the degree of substitution. Fig.6 shows the current densities at constant electrode potential as a function of the degree of substitution of cobalt and the function described by eq. (10) calculated for the ideal case. The experimental current densities are much higher than the calculated function.

The same results were obtained for all the investigated systems. As shown in Fig.7, most of the substituents decreased the dissolution rate of the iron-based alloys. Manganese was the only exception. The following general rule emerges: If in an alloy corroding in the active state a component with a high rate constant is substituted by another component with a low rate constant, the dissolution rates at intermediate degrees of substitution are larger than those expected for an ideal systems. This rule applies both for the glassy and the crystalline alloys.

The deviations from ideal behaviour are much greater than expected from the estimated concentration dependence of the bulk activity coefficients. It follows from the experiments that at intermediate degrees of substitution the small rate constant appears to be larger and the high rate constant smaller than in the ideal case. The most probable explanation is as follows: At the surface of a crystal dissolution proceeds in a repeatable manner at kinks in steps. If a kink position is occupied by an atom of the slowly dissolving component, this position will be blocked for some time. However, at a given electrode potential, the component with the faster rate constant will have a fair chance to dissolve from a position energetically less favourable than a kink, e.g. from a position on a fully occupied step. Thereby new kink positions are created which after some elementary acts will again be occupied, mainly by the component with the slow rate constant. As a result, due to the large surface concentration of kinks, the apparent rate constant of the slowly dissolving component is increased. On the other hand, the apparent rate constant of the quickly dissolving component is decreased, because it is forced to dissolve from the energetically less favourable positions. The result can also be described as roughening on an atomic scale.

At the surface of glass, well defined kinks and steps do not exist, but there is an almost continuous spectrum of positions with different binding forces. Again, one expects that in the steady state the majority of positions with small binding force are occupied by the slowly dissolving component and those with high binding force by the quickly dissolving component. The effect on the apparent rate constants is qualitatively the same for the glass and for the crystal.

Substitution of chromium for iron in glassy FeCrB alloys also lowers the dissolution rates in the active state. At degrees of substitution in excess of about z=0.1 the passivating current density decreases and the steady state corrosion rate in the passive state begins to drop.

Chromium is the only boron-containing glass which exhibits normal passivity. With all the other metal-boron glasses very unstable passivity was observed with high corrosion rates in the passive state. The behaviour often was typical of passivation by some salt layer and not by a compact oxide film. The unstable passivity results from the high boron content and not from the glassy structure.

REFERENCES

1) Kapusta, S.D. and Heusler, K.E.: Zeitschr. f. Metallkunde 72, 785 (1981).
2) Huerta, D. and Heusler, K.E.: J. Noncryst. Solids 56, 261 (1983).
3) Huerta, D. and Heusler, K.E.: Proc. 9th Intern. Congr. Metallic Corrosion, Vol. I, 222, Toronto 1984.
4) Heusler, K.E.: The Electrochemistry of Iron; in: Encyclopedia of the Electrochemistry of the Elements, Vol. IXA, 299 ff., ed. by Bard, A.J., Dekker, M., Inc., New York and Basel 1982.
5) Folleher, B. and Heusler, K.E., J. Electroanal. Chem. 180, 77 (1984).

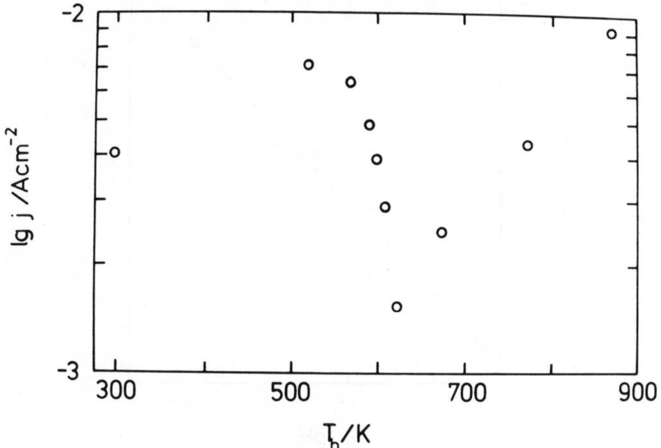

Fig.1. Dissolution rates of $Co_{75}B_{25}$ in 0.5 M sodium sulphate solution, pH 1.8, at E = -0.3 V vs. SCE and 298 K after heat treatment for 5 h at the temperature T.

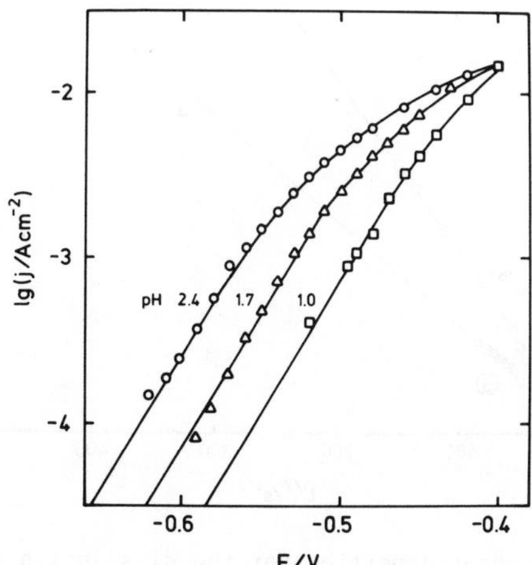

Fig.2. Steady state polarization curves of $Fe_{80}B_{20}$ glass in 0.5 M sodium sulphate/sulphuric acid solutions at 298 K.

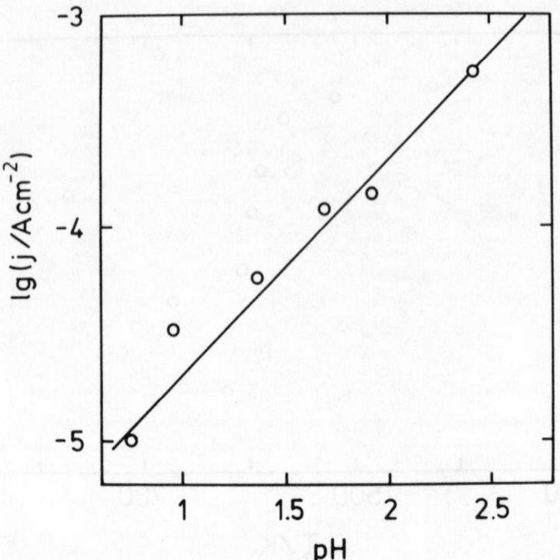

Fig.3. Steady state current densities for dissolution of $Fe_{80}B_{20}$ in 0.5 M sulphate solution at E = -0.582 V vs. SCE as a function of the pH value.

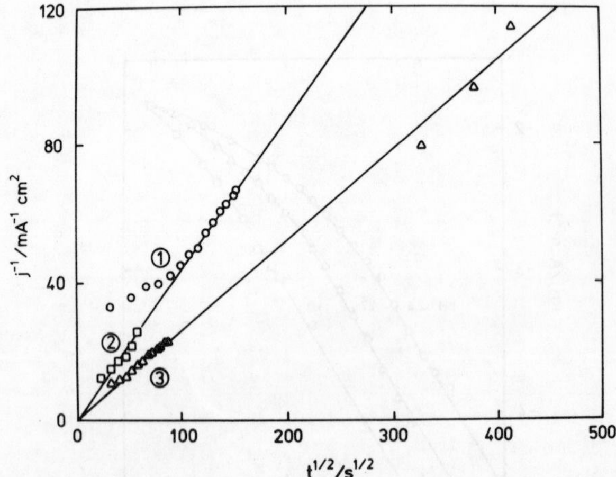

Fig.4. Reciprocal current densities for the dissolution of boron from $Pt_{60}B_{40}$ in (1) 0.5 M sodium sulphate solution, pH 1.8, at E = -0.36 V vs.SCE; in 1 M sodium carbonate/bicarbonate buffer solution, pH 9.0, (2) at E = -0.48 V vs. SCE and (3) at E = -0.744 V vs. SCE.

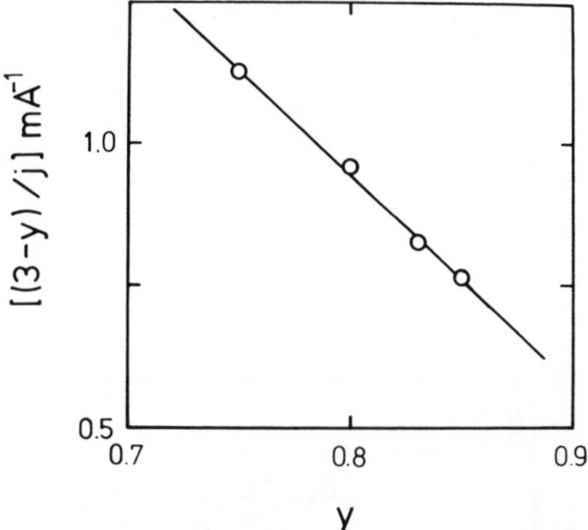

Fig.5. Relation between the reciprocal current density for steady state dissolution of Fe_yB_{1-y} glasses vs. molar fraction y, according to eq. (8).

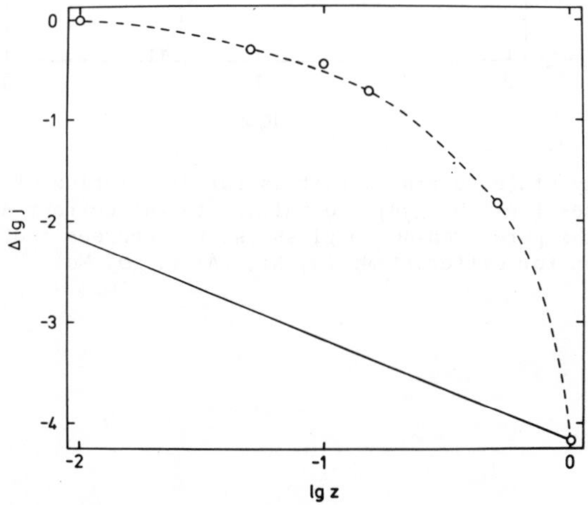

Fig.6. Steady state current densities j for the dissolution of active glasses $(Fe_{1-z}Co_z)_{87}B_{13}$ normalized to the current density for the pure iron-boron glass vs. the degree of substitution z.

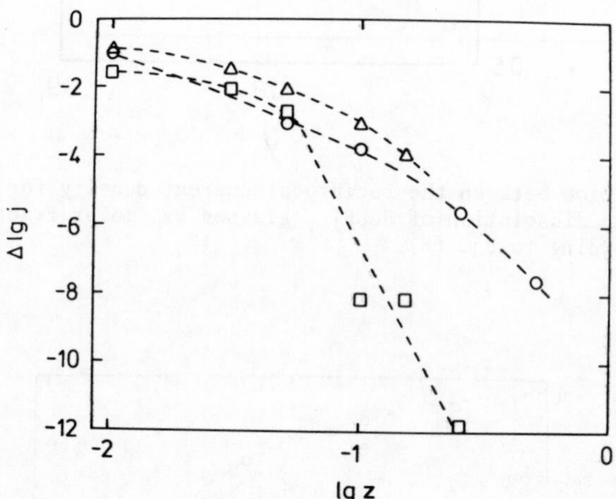

Fig.7. Steady state current densities for dissolution of active glasses $(Fe_{1-z}M_z)_{87}B_{13}$ normalized to the current density for the pure iron-boron glass vs. the degrees z of substitution for different M: (o) Ni, (Δ) V, (□) Mo.

APPLICATIONS OF METALLIC GLASSES

Henryk K. Lachowicz

Institute of Physics, Polish Academy of Sciences,
al. Lotnikow 32/46, 02-668 Warszawa, Poland.

1. INTRODUCTION

Metallic glasses combine excellent soft magnetic properties and extraordinarily good mechanical behaviours, due to the lack of magneto-crystalline anisotropy and the absence of dislocations and their movement, respectively. It is worth noting that such a combination of properties is not found in any of the known crystalline magnetic materials.

An additional advantage of metallic glasses is the ease of selection of their composition. Due to this, metallic glasses can easily be tailored according to the requirements of a given application.

Because of their behaviour, metallic glasses can be utilized in almost every branch of industry (see e.g. [1,2]). Since the number of possible applications is rather large, the present survey will be limited to major magnetic utilizations of the alloys considered.

Moreover, only these applications will be considered which utilize metallic glasses fabricated in the form of ribbons. A number of possible applications of metallic amorphous alloys prepared as sputtered or thermally deposited films will not be presented here, because of space limitations. The Reader who is interested in these applications, will find a comprehensive description of this topic in a number of published surveys and monographs (e.g. [3,4,5]).

It is the author's hope that the choice performed will meet the expectations of the majority of the participants of this School.

2. MAJOR MAGNETIC APPLICATIONS OF METALLIC GLASSES

The major magnetic applications of metallic glasses can generally be found in one of the following three categories:

i) Power electric applications,

ii) Transducers and sensors,

iii) Magnetic shielding.

The majority of the devices which can be classified as belonging to one of the categories listed above, will be presented with the emphasis on the advantages gained as a result of application of the metallic glasses.

2.1. Power Electric Applications

Among all the devices in this category, the most intense research efforts are devoted to the distribution transformers, since application of metallic glasses as the magnetic core material in these devices should enable significant energy savings.

On the other hand power transformers seem to be the most important utilization of metallic glasses in future, since they will consume the great majority of the demand for these materials.

Apart from the distribution transformers, there are a number of other possible applications which, though not consuming as much volume of metallic glasses ribbons, are nevertheless worth presenting.

2.1.1. Distribution transformers. A remarkable feature of the distribution transformers, one which distinguishes them from all the other transformers, is that they are permanently energized during their exploitation time, usually over 30÷40 years. As a result, the energy supplied by the utility is wasted due to magnetic loss in the transformer core cycled around its B-H loop at the line frequency. To convey an idea of the magnitude of this energy loss Raskin and Smith [6] gave an impressive example of a 25 kVA conventional core transformer, which operates at 98.7% efficiency at full load, but generates 85W of core loss independent of loading and this, during the 35 years of service, wastes an energy of more than 26 x 10^3kVh!

Since the metallic glasses exhibit a considerably lower loss, on average 65% to 75% lower than that of the grain-oriented Si-steel, a significant amount of the wasted energy could be saved if using these materials.

The condition of a low power loss, however important, is not the only requirement for a core material in distribution transformers. For this application, one needs a material which meets the following requirements (see e.g. [7]):

- high saturation induction,
- low core loss,
- low exciting current (current required to maintain the chosen induction level),
- low magnetostriction,
- long-term durability,
- large stacking factor.

The only metallic glasses acceptable for this purpose are the Fe-based alloys, because of their relatively low costs. Though, in general, they exhibit lower saturation inductions (e.g. B_s = 1.56 T for the 2605 S2 Metglas*) than that of Si-steel (B_s = 2.03 T for Fe-3% Si), nowadays steadily rising energy costs reduce this disadvantage because of the existing tendency, even for Si-steel transformers, to operate at lower induction levels, to decrease the core loss (eddy current losses are proportional to the induction squared).

As it has been mentioned above, the extremely low power losses of the Fe-based metallic glasses are the main feature which makes them very attractive for utilization in distribution transformers. Fig. 1 shows the loss characteristics of the Fe-based metallic glasses fabricated by different manufacturers and, for comparison, also the one of Fe-3% Si.

---

* A registered trademark for metallic glasses manufactured by the Allied Corporation.

It is well known that the power loss in any magnetic material is composed, in general, of two contributions. The first expresses the static hysteresis loss, a measure of which is the area enclosed by the static B-H loop, whereas the second one is due to the eddy currents induced in the magnetized material. The static hysteresis loss is comparatively low in metallic glasses, since they show extremely small coercivity, due to the lack of magnetocrystalline anisotropy. Since the electrical resistivity of metallic glasses is several times larger than that of crystalline materials, one should expect a reduction of the eddy current loss by the same amount. In general, this is not the case. The eddy current loss contribution is usually considerably larger than the sum of the measured static hysteresis loss and the eddy current loss, calculated according to the classical formula derived from Maxwell's equations under the assumption of uniform flux penetration (see e.g. [8]). This, the so-called excess loss, is usually responsible for as much as 90% to 99% of the total power loss of metallic glasses at power frequencies (see e.g. [9]). The excess loss is mainly caused by the eddy currents induced by the moving domain walls. It is known that this loss scales with the ratio of the domain wall spacing to the ribbon (or lamination) thickness [10]. Metallic glasses tailored for distribution transformer cores exhibit uniaxial anisotropy aligned with the ribbon length, induced by means of a thermomagnetic treatment of the as-quenched ribbons. As a result, the magnetic structure consists of rather coarse bar-type domains, directed along the ribbon axis. Since the ribbon thickness is small, the scaling factor mentioned above is relatively large. To reduce it, one should refine the domain structure.

The only efficient method of domain refinement seems to be the controlled generation of locally crystallized regions. This can be done by proper annealing (e.g. [11,12]) or by a laser treatment of the ribbon [13]. Application of such a procedure results in reduction of the power loss by as much as 30 per cent [13].

In particular, the requirement of a low exciting current can easily be fulfilled in the case of metallic glasses at relatively low induction (typically this current is approximately one-third that in a conventional Si-steel [1]).

Fe-based metallic glasses exhibit large magnetostriction (e.g. 27×10^{-6} for Metglas 2605 SC), this being undersirable behaviour since it gives rise to an audible noise with the second harmonic of the operating frequency. Additionally, because of large magnetostriction, the magnetic properties of the ribbon should deteriorate at the time of core forming via magnetoelastic interactions.

However, Datta et al. [14] have shown that since the magnetostrictive strain responsible for transformer acoustic noise is the net strain parallel to the applied field (called "engineering magnetostriction"), for magnetic glasses, as well as for grain-oriented Si-steel, this strain is very small, since the direction of the induced uniaxial anisotropy is aligned with the applied field direction. Consequently, the undesirable influence of magnetostriction on the metallic glass transformer parameters should be very similar to that in conventional grain-oriented Si-steel.

As regards the long-term durability of metallic glasses used in distribution transformers, thermal stability is the most important factor. This can be estimated by measuring the crystallization temperature. However, it is rather difficult to define the service life of a transformer exclusively on the basis of this estimate since magnetic properties deteriorate with time at the temperature much lower than the crystallization temperature. For more accurate evaluation of the service life, one should carry out the so-called accelerated ageing experiment, in which the selected magnetic parameters are measured at temperatures higher than the operation temperature of the transformer. An extrapolation of the data obtained to the actual operating temperature allows us to estimate the service life of the transformer. Such measurements, performed by Datta et al. [15] for the 2605 S-2 Metglas, have shown that only the last one of three chosen parameters, namely: power loss, coercivity and exciting current, should display detectable changes after 50 years of service.

Accelerated ageing experiments showed that the observed deterioration of magnetic properties is mainly due to surface oxidation of the ribbon [15,16].

The occurrence of this effect forces one to search for a protective coating. On the other hand, a protective layer should further reduce the effective space factor which, for metallic glass ribbons, is much smaller than that for conventional materials. Moreover, such a layer should introduce an undesirable stress. For all these reasons (as well as to eliminate additional manufacturing processes), protective coating is not usually applied for ribbons used in distribution transformer cores.

The relatively small value of the stacking factor for a metallic glass ribbon, ranging from 75% to 85%, as compared to 95% or even better in the 300 μm thick Si-steel, is mainly due to the thinness of the ribbon (usually 25 to 40 μm thick). For a given thickness of core winding one needs the number of layers which is larger by almost an order of magnitude than that required if the core is laminated using conventional materials. Since ribbons usually exhibit surface irregularities, the increased number of layers enlarges their influence on the effective space factor.

However, though ribbons as thick as 250 μm can be obtained in an amorphous state (e.g. [17]), nevertheless, there have not been any reports indicating that such a thick ribbon can be produced with a width which is adequate for practical uses.

As a result of the small stacking factor, the dimensions of the transformer core made by utilizing metallic glasses are larger than those of a conventional one.

Considering the above description, it is evident that metallic glasses do not meet all the requirements of the material to be used in a distribution transformer core. Additionally, because of the ribbon's geometry, conventional, well-established design and manufacturing techniques are, in general, not applicable for metallic glasses. The approach most frequently applied commercially is the so-called cruciform-core transformer, constructed using ribbons of different widths, wound so that the cross-section of the core approximates a circle (e.g. [18]).

None the less, metallic glasses seem to be the material for distribution transformer-cores which will compete with the conventional ones, mainly because the energy saving factor becomes more and more important.

The visible sign of this challange are the efforts carried on, particularly in the United States and Japan, yielding successful transformer constructions of from 10 kVA up to 100 kVA power, being now tested by the various utility companies (see e.g. [19,20]). In Europe, because of the different electric energy distribution system (electricity with a common voltage is provided centrally) this research does not seem to be so intense.

2.1.2 400 Hz power transformers. Metallic glasses can be successfully utilized also as the core material for 400 Hz power transformers, commonly used for electric energy distribution on board of aircrafts and ships as well as for the military hardware.

To show the advantage of the use of the metallic glass core, one usually presents the most impressive comparison, originally produced by Boll and Warlimont [21], which shows the dependence of the power output versus the peak induction of a 400 Hz transformer with a torroidal strip-wound core constrained to the maximum temperature rise of 75 K above the 40°C ambient. Their results, as shown in Fig.2, indicate that the power output can be raised by 20 or 60 per cent for a Metglas 2605 SC core as compared to the Si-steel core made using strips 0.1 and 0.3 mm thick, respectively.

Instead of increasing the power one should, equivalently, reduce the core weight and volume, a measure which should be highly velcome to the aerospace industry.

The results shown in Fig.2 should even be better when applying a metallic glass especially tailored for this application [22], namely Metglas 2605 CO ($Fe_{67}Co_{18}B_{14}Si_{1}$) which exhibits a larger saturation induction (1.8 T) than that of Metglas 2605 SC (1.6 T). Though the former shows a relatively large loss (approximately 3 times larger than that of Metglas 2605 SC), this factor seems nevertheless to be not as critical as the volume and weight of the device, which should be smaller and lighter if the operating induction level have to be higher.

It seems that the only reason for which the 400 Hz metallic glass transformers are not yet in serial production is the insufficient number of reliable tests performed up till now.

2.1.3 Pulse power devices. The next possible application of metallic glasses which can also be included in the category discussed here, is in the torroids utilized in linear induction accelerators (linacs for short). These torroids, surrounding the beam tube, act as 1:1 transformers, supplying the acceleration voltage appearing at the secondary winding as a result of a short, high energy pulse applied to the primary one. Since this voltage is proportional to the time derivative of the induction change, the maximum possible swing, from minus remanence to plus saturation, is needed to obtain the highest voltage or to minimize at the given voltage and pulse width the volume of the material and, consequently, its cost, because of the large quantity involved (typically 100 torroids, each 50 to 500 kG in weight [1]). Usually, Si--steel, 50% Permalloy and ferrites are used as the torroid material, depending mainly on the applied pulse width. However, it was recently shown that the Fe-based metallic glasses can offer a significant reduction of the losses in the pulse width range of 1-10 μs, and that they can also be efficient even for very short pulses (0,05 μs) if the ribbon thickness is reduced up to ∿ 15 μm [23].

The availability of metallic glass ribbons as thin as 15 μm has led to a number of applications in the area of the pulse power technique, apart from that in the particle accelerator drivers already described. For example, one can mention power laser modulators, as well as saturable reactors which serve as pulse compressors (e.g. [18,24]). The latter are closely related to the liniacs, since they can be used to compress high energy pulses to the required width.

To compare metallic glasses with conventional materials from the standpoint of pulse power applications one can use a figure of merit (FOM) which combines the resistivity ρ, thickness d and the induction swing ΔB (from negative remanence, B_r, to plus saturation, $+B_s$)

$$FOM = \frac{\Delta B^2 \cdot \rho}{d^2} \quad . \tag{1}$$

It is worth noting that similar combinations of the parameters used in the above definition enter into the values of the magnetic losses, unsaturated permeability and also into calculations of the core volume [22].

In Table 1, the calculated values of FOM are presented for the Metglas 2605 CO mentioned earlier, which has also been tailored for pulse-power applications [22] (as well as for 400 Hz power transformers), and compared with the well-known Metglas 2605 S-2 designed for an extremely low core loss, as well as with conventional policrystalline materials.

The comparison presented in Table 1 shows that the best choice for power-pulse applications is the 2605 CO Metglas (the FOM for this alloy should even be larger, by a factor of 2.8, for a ribbon 15 μm thick). The limitations on a wider utilization of this material seem to be its present cost as well as some technological problems arising when a core is being formed.

2.1.4 Switched mode power supplies. Magnetic components for switched mode power supplies (SMPS) can also be classified in the category of power electric applications. The market for these components is rapidly growing, since the SMPS are widely replacing the less efficient standard mains power supplies in many electronic applications. The operating frequency of the SMPS is significantly higher and corresponds to that of fast VMOS transistors (see e.g. [25]), being in the range of 20 to 100 kHz. The variety of the inductive components used in SMPS circuitry is clearly seen in Fig.3, which shows a typical diagram of a multioutput device. For frequences higher than 20 kHz, the primary requirements for the core material of the magnetic components are a low power loss and a high saturation induction. Metallic glasses, in particular those of the highest quality (Co-based alloys), meet these requirements. Moreover, their hysteresis loop can easily be tailored to the shape required for the particular component by subjecting the ribbon to a proper thermomagnetic treatment (see e.g. [26]). These features permit one to construct practically all the inductive components of a SMPS from metallic glasses, so as to yield a higher performance, lower loss, smaller dimensions and, in a number of cases, even a cost lower than that of the same components built by using conventional materials [27]. A comparison of the metallic glasses with Permalloys and ferrites, used to date in the SMPS, shows that they are comparable with thin Permalloys, but are superior to the ferrites for frequencies of up to 100 kHz and even above [28]. Some of the metallic glass components for the SMPS are now commercially produced (for example, by the Vacuumschelze GmbH). It is worth noting that a SMPS utilizing only metallic glass core components has been constructed for up to 1 kW output power [29].

2.1.5 Electric motors. Apart from the distribution transformers, electric motors are, by number and universality of utilization, the devices which consume the greatest volume of magnetic materials produced nowadays. Since the specific magnetic losses in motors are usually much higher than in power transformers, the efficiency of these devices is lower (~85%), and it is not surprising that motors account for half the global energy loss in electric machinery.

Whereas in transformers the applied flux is uniaxial and the direction of the magnetic anisotropy of the core material can be aligned with the flux direction using grain-oriented Si-steel, in motors such a situation does not arise. The flux path is generally nonuniform and, for that reason the, non-oriented Si-steel is commonly used, leading to an increase in the power loss.

Metallic glasses give a unique opportunity to improve the efficiency of electronic motors, since their anisotropy can be "built-in" during the final annealing, performed in such a manner that the direction of the anisotropy will follow exactly the flux path configuration required by the particular motor design [1].

As a result of the application of metallic glasses, power losses in motors can be dramatically reduced by as much as 90 per cent of these corresponding to the situation when a conventional material is used [1].

However, wider utilization of metallic glasses in electric motors will be possible after overcoming some serious manufacturing problems, arising because of the small ribbon thickness and the fact that it is very hard.

Furthermore, the Fe-6.5% Si rapidly solidified crystalline alloy [30], seems to be a competitive material for this application, since it is magnetically isotropic in-plane, exhibits high saturation induction, relatively low losses and "zero" magnetostriction behaviour. Moreover, its low hardness allows one to use conventional fabrication techniques.

2.2 Transducers and Sensors

This category comprises a variety of devices which, unlike the former class, require an application of only grams of metallic glass ribbons. The category includes some already mass-produced devices, like audio and video magnetic heads, as well as a number of novel devices, such as sensors of non-electric quantities or acoustic delay lines tunable by an external magnetic field, both utilizing the extraordinary magnetoelastic properties of the metallic glasses. As in the case of the magnetic heads, in which a replacement of conventional magnetic materials by metallic glasses significantly improves the functional quality of the device, also other devices are considered in this category such as inductive press keys, thermo-switches or anti-pilferage devices, which operate better as a result of application of metallic glasses. These applications will be briefly described with particular attention being paid to the devices designed and constructed by the author's colleagues from the research centers collaborating in the projects devoted to amorphous materials.

2.2.1 <u>Magnetic heads</u>. Metallic glass magnetic heads (audio, in the beginning) have been successfully introduced commercially as one of the earliest applications of these materials.

Materials for magnetic heads, based on the induction principle and used in contemporary audio, video and data storage systems, require a combination of superior electromagnetic and mechanical behaviours. The most important of the electromagnetic properties are the following (see e.g. [31]):

- high saturation magnetization, since the common use of highly coercive magnetic storage media (tapes, disks) means that a high flux density of the head is reguired to achieve high reliability and high recording fidelity,
- high permeability, in order to achieve high recording and playback efficiency of the head, owing to reduction of the magnetic flux leakage,
- low saturation magnetostriction, to protect against deterioration of the magnetic properties of the material which should occur via the magnetoelastic interaction in the head forming process and also as a result of the mechanical vibrations and shocks due to the moving storage medium,
- low losses, i.e. high electrical resistivity and low coercivity, the latter also important in protecting against the noises introduced by the head during playback.

Since the moving storage medium usually heats up the head as a result of mechanical friction, sufficient thermal stability of the properties is also desired.

The most important mechanical properties required are the high wear resistance, hardness and stiffness. The first of them is one of the conditions which determines the life time of the head.

Non-magnetostrictive (usually $|\lambda_s|<10^{-6}$) Co-based glasses meet practically all the requirements listed above. The only disadvantage of these alloys is that all the head forming processes should be performed below the crystallization temperature, which is of the order of 400 to 500°C (for conventional materials, since they are crystalline, this limitation does not exist). In order to avoid possible crystallization, the head core lamination process should be performed using laser spot welding (the procedure applied by Sony Co. in commercial production of audio heads, hence called "laser amorphous heads" [31]. Whereas the audio and video metallic glass heads are already mass-produced (mainly in Japan), data heads for digital applications which utilize amorphous alloys are now under trial production.

Properties of metallic glasses especially suited to head core applications can be exemplified by the AMM-metallic glass (a Co-based alloy designed and manufactured by the Institute of Material Science and Engineering of Warsaw University of Technology). This alloy, after annealing at 430°C/20 min, has the following characteristics:

- saturation induction (at 1.6 kA/m)	$B_s \geqslant 0.8$ T
- coercivity	$H_c \leqslant 1.6$ A/m
- permeability ($\mu_{0.4}$ at 1 kHz)	$\mu_{0.4} \geqslant 14000$
- maximum permeability	$\mu_{max} \geqslant 90000$
- drop in permeability after ageing at 55°C/750 h	$\mu_{0.4} \leqslant 16\%$
- saturation magnetostriction	$\lvert\lambda_s\rvert \leqslant 1\times10^{-6}$
- Curie temperature	$T_c = (420\pm5/^{o}C$
- crystallization temperature	$T_x = (470\pm5)^{o}C$
- electrical resistivity	1.15 μΩm
- density	$d = 8.0$ g/cm^3
- hardness	$HV_{100} = 1000$-1100
- yield strength	$R_e = 1.4$ GPa
- ribbon thickness (max)	$t = 40$ μm
- ribbon width	up to 60 mm

The frequency dependence of the permeability and the power loss vs. peak induction, measured for this alloy, are shown in Figs. 4 and 5, respectively.

The AMM-metallic glass has been utilized in trial production of magnetic heads. The tests performed showed that this material can be successfully applied in laminated head-cores for high-fidelity consumer products as well as in professional aquipments [32].

Since the Co-based metallic glasses are relatively expansive (with prices ∿ 30 per cent higher than these of Fe-based alloys), attempts have been made to utilize the Fe-based glasses in audio-heads for low class tape recorders [31].

Some attempts have also been made to increase wear and corrosion resistance. For example, Sakakima et al. [33] showed that adding niobium to a Co-based glass improved wear resistance more than five times, whereas addition of chromium, as it is well known, increases corrosion refractoriness. In the latter case, a protective layer of chromium oxide develops on the ribbon surface, minimizing oxidation of the metalloids [34]. An additional advantage is an increase of the interlaminar resistance, leading to lower eddy current losses.

It seems that, in future, metallic glasses should force out classical materials, such as Permalloys, ferrites, and even Sendust (Fe-Al-Si alloy) from the magnetic head market.

2.2.2 Sensors of non-electric quantities. Metallic glasses, with their superior magnetomechanical properties, have given a new impulse to the development of sensors operated on the magneto-elastic principle. Since in these materials the magnetic anisotropies should be vanishingly small, mechanical stresses strongly influence the permeability as well as the hysteresis loop via the induced magnetoelastic anisotropy, proportional to the product of the saturation magnetostriction and the applied stress.

The class of sensors considered here comprises a great variety of devices used for detection and/or measurement of non-electric quantities, for example: displacement, distance, pressure, shock, force, angular velocity, torque, touch, heat.

By considering the type of metallic glass applied, these sensors can, in general, be divided into two groups: those devices utilizing non-magnetostrictive and those utilizing magnetostrictive alloys.

Since all of these sensors, as well as those for sensing electric quantities (e.g. electric current), have been excelently reviewed in a series of papers presented by Mohri (see e.g. [35,36]), in the present survey, only the mechanical torque and angular velocity sensors will be presented. The first one, developed by Harada, Sasada and co-workers [37,38,39], offers unique opportunity of a contactless, continuous detection and measurement of the instantaneous torque and should be utilized wherever a torque has to be controlled, e.g. in machine tools. The second one, developed to measure the angular velocity of a Diesel engine [49], is worth presenting, because of its principle of operation, which seems to be quite ingenious. Both of them fall into the group of sensors utilizing the magnetostrictive metallic glasses.

The sensing elements in the torque detecting device are pairs of elongated rectangles (or parallelograms), cut from the Fe-based glass ribbon and attached onto the surface of rotating shaft, in which the torque has to be controlled. These rectangles are located so as to form an angle of $\pm 45^{\circ}$ with the shaft axis (see Fig.6). The magnetization in each of the rectangles is aligned with the $\pm 45^{\circ}$ directions because of shape anisotropy. The sensing elements are magnetized by the sinefield generated by a solenoid (see Fig.6). Under the non-load condition (no applied torque) the voltage induced in each of the two pick-up coils is of the same amplitude. Since the coils are connected in series, but in opposite phase, the output voltage is zero. An application of a torque creates, via the magnetoelastic interaction, additional anisotropy. For the left-hand rectangles (see Fig.6) this anisotropy is directed along their width, whereas for right-hand ones, along their length, or vice versa, depending on the signs of the torque and the magnetostriction (application of the torque induces orthogonal compressive and tensile stresses in the $\pm 45^{\circ}$ directions, as seen in Fig.6). Because of the opposite directions of rotation of the axis of effective anisotropy, due to the effect mentioned above, the axial component of magnetization increases in the left and decreases in the right-hand rectangles, respectively (or vice versa). As a result, the amplitude of the pick-up voltages induced in each coil because unbalanced, and the output voltage has a non-zero value which is the larger, the larger is the applied torque.

The dependence of the output voltage on the applied torque is shown in Fig.7 for the sensor's parameters as indicated.

It has been shown that metallic glasses can be used for sensors of angular velocity based on the large Barkhausen jump [41] and the Matteucci [42] effects. Operation of these devices requires access to the rotating element, the velocity of which has to be measured. In some cases, this requirement can be inconvenient, e.g. combustion engines.

In Diesel engines, a sudden change of the pressure in the fuel pipe occurs at the moment of injection. This effect can be utilized to measure the angular velocity, eliminating the requirement mentioned above. The construction of the sensor based on this effect is shown schematically in Fig.8. A magnetostrictive metallic glass ribbon is wound helically onto the fuel pipe and attached firmly to its outer surface, so that the strain in the pipe should be transferred into the ribbon. The ribbon has no galvanic contact with the pipe surface. A solenoid which produces an axial d.c.-field is located coaxially with respect to the pipe. As a consequence of the Matteucci effect, a sharp voltage pulse appears between the ends of the ribbon at the moment of fuel injection. The number of pulses counted in a time unit gives a measure of the angular velocity of the Diesel engine.

Fig.9 shows a typical output pulse obtained in an experimental device designed and constructed in the Institute of Theoretical Electrotechnique and Electrical Measurements of the Warsaw Technical University [40]. The pulse shown in this figure appears in the case of the $Fe_{78}B_{13}Si_9$ metallic glass ribbon (1.25 mm wide and 42 μm thick) wound onto a fuel pipe of 6 mm diameter and for the change of pressure of 6 MPa (a pressure of 200 MPa occurs at the moment of injection in some types of Diesel engines).

2.2.4 Other devices. There are quite a number of other devices using metallic glasses which could be classified in the category of sensors and transducers.

Among all these devices there are a few which in the author's opinion, are worth noting, namely: tunable delay lines which are one of the first practical uses of magnetostrictive metallic glasses, phonograph cartridges-devices mass-produced by Sony in Japan [43], and the anti-theft tags made from metallic glass ribbons to protect expensive articles in department stores, as well as items in libraries (see e.g. [18]).

Fe-based metallic glasses show a giant ΔE-effect, owing to their outstanding magnetomechanical properties (e.g. for Metglas 2605 SC estimates give a magnetomechanical coupling constant of 0.98 and $\Delta E/E$ exceeding ten[44]).

Since the velocity of an acoustic wave propagating in a medium of density ρ and Young's modulus E, is proportional $(E/\rho)^{1/2}$, magnetostrictive metallic glasses can serve as an excellent material for delay lines tunable by an external magnetic field. It has been verified by Arai and Tsuya [45] that delay lines can successfully operate on this principle.

2.3 Magnetic Shielding

Historically, the first commercial application of the ferromagnetic properties of metallic glasses was in magnetic shields, manufactured in the form of a flexible woven or braided mesh from a narrow ribbon (2 mm-wide) of the NiFePB metallic glass (Metglas 2826)[46].

Magnetic shielding is particularly required for static and low frequency disturbing fields. In this case, the usual conductive shields are not effective since the penetration depth of an a.c.-field scales with $1/(f)^{1/2}$, where f is the field frequency. The shielding factor, which is the ratio of the magnetic field outside to that inside the shield, depends on the geometry and is the larger, the higher the permeability of the shield material. Therefore, for effective shielding, the permeability should be as high as possible. Metallic glasses, especially those exhibiting low magnetostriction, are very suitable for this purpose. In addition to their high permeability, (e.g.: $\mu_{max} \simeq 600000$ for Vitrovac 6025*)), shields made from these materials can be deformed mechanically in manufacturing and in service, without deterioration of their magnetic properties.

The superiority of metallic glass screen is seen in Fig.10, in which a shielding factor versus the intensity of the disturbing field

---

*) A registered trademark for metallic glasses manufactured by the Vacuumschmelze GmbH.

is given for Co-based alloy and steel shields, made in the form of a cylinder of 30 mm diameter, 150 mm length and about 0.3 mm wall thickness. In the case of metallic glass ribbon this cylinder was prepared by winding the appropriate number of isolated layers so as to obtain the same wall thickness [47].

Insensitivity of the Co-based metallic glasses to bending gives a unique opportunity to utilize them for cable shielding. This application has been demonstrated long ago [48], but the ribbon was then braided in the way it was for a conductive shield. Because of the numerous air gaps which limit the effective permeability and thus the shielding factor, it has been shown that better results should be obtained if the ribbon were to be wound helically around the cable [49]. However, Hilzinger [50] has shown that the shielding may be further improved by adding a second layer wound in the opposite sense. The experiments performed (using Vitrovac 6025 X, the Co-based alloy) showed that the shielding factor is of the order of 100 for a spacing of up to 1 mm between the turns of the ribbon, a value which is larger by a factor of 8 as compared to cable-shielding using a Permalloy ribbon [50].

Application of metallic glass shields on board of space ships of the Voyager missions seems to be a good recommendation of the screening behaviour of these materials.

3. CONCLUSION

Because of their curious properties, metallic glasses, particularly those which are magnetically ordered, have for a long time been laboratory objects tortured by scientists who wanted fo find the answers to many questions arising from the unexpected experimental results obtained.

Though not all of these questions have yet been answered, metallic glasses have, nevertheless, moved from the laboratory to applications in commercial products, as one can observe even on the basis of the present survey.

Wider utilization of these materials requires however, further price reduction for it to become cost effective when used in a larger range of applications, and what seems to be the most important, a rejection of the present strong effection for the old, well established design methods of devices tailored to conventional materials.

The energy savings problem, which becomes more and more important, also seems to be a crucial factor in the challange the metallic glasses undertook.

Finally, the dramatic words of Gonser [51)] should be cited: "we have reached (in metallic glass R&D) a crossroads where amorphous metals are either going to become the material of the century or just turn out to be a dream".

It is the author's belief that the first of these possibilities will be realized.

REFERENCES

1) Raskin, D. and Smith, C.H., "Amorphous Metallic Glasses, ed. F.E. Luborsky, Butterworths, chapter 20, London 1983, p.381.
2) Taub, A.I., Proc. V Int.Conf. on Rapidly Quenched Metals, eds. S. Steeb and H. Warlimont, vol.II,p.1611, North-Holland, 1985.
3) Lachowicz, H.K., J. de Physique, suppl. nr C6, 46, C6-181 (1985).
4) Moorjani, K. and Coey, J.M.D., "Magnetic Glasses", North-Holland, 1984.
5) Buschow, K.H.J., "Handbook on the Physics and Chemistry of Rare Earths", eds. K.A. Gschneider Jr. and L. Eyring, Elsevier Science Publishers B.V. 1984.
6) As the ref. 1, p.382.
7) Sato, T. "Recent Magnetics for Electronics", ed. Y.Sakurai, Ohmasha Ltd. and North-Holland, chapter 13, p.151, 1983.
8) Heck, C., "Magnetic Materials and their Applications", Butterworths, London 1974.
9) Overshott, K.J., IEEE Trans. Magn., MAG-17, 2698 (1981).
10) Pry, R.H. and Bean, C.P., J. Appl. Phys., 29, 532 (1958).
11) Hasegawa, R., Fish, G.E. and Ramanan, V.R.V., Proc.IV Int.Conf.on Rapidly Quenched Metals,eds. T.Masumoto and K.Suzuki, Jpn.Inst. of

Metals, Sendai, vol. II, p.929, 1982.

12) Kan, T., Shishido, H. and Ito, Y. 88th Meeting of Japan Institute of Metals, Tokyo, 1981.
13) Sato, T., Yamada, T. and Ozawa, T., ref. 2, p.1643.
14) Datta, A., Nathasingh, D., Martis, R.J., Flanders, P.J. and Graham, C.D., Jr., J. Appl. Phys., 55, 1784 (1984).
15) Datta, A., Martis, R.J. and Das, S.K., IEEE Trans. Magn., MAG-18, 1391 (1982).
16) Sato, T., Ozawa, T. and Nagumo, M., ref. 11, p.961.
17) Hagiwara, M., Inoue, A. and Masumoto, T., Sci. Rep., RITU, A-29, 351 (1981).
18) Smith, C.H., IEEE Trans. Magn., MAG-18, 1376 (1982).
19) Bailey, D.J., and Lowdermilk, L.A., ref. 2, p.1625.
20) Yamamoto, Y., ref. 2, p.1629.
21) Boll, R. and Warlimont, H., IEEE Trans. Magn., MAG-17, 3053 (1981).
22) Datta, A. and Smith, C.H., ref. 2, p. 1315.
23) Schwarzschild, B.M., Physics today, 35, 20 (1982).
24) Hasegawa, R., J. Non-Cryst. Sol., 61-62 (1984).
25) Kunz, W. and Grätzer, D., J. Magn. Magn. Mat., 19, 183 (1980).
26) VITROVAC-Amorphous Metals, Vacuumschmelze GmbH, Edition 2/83,Fig.6.
27) Warlimont, H., Helv. Phys. Acta, 56, 281 (1983).
28) Warlimont, H. and Boll, R., J. Magn. Magn. Mat., 26, 97 (1982).
29) Torre, J.J., Smith, C.H. and Rosen, M., Proc. 4th Int. Power Conversion Conf., San Francisco, p.278, March 1982.
30) Das, S.K., DeCristofaro, N.J. and Davis, L.A., ref.2, p.1621.
31) Makino, Y., ref. 2, p.1699.
32) Kędziorek, H.,"Construction of magnetic heads", Report of the R&D Center for the Broadcasting and TV Techniques, Warszawa, April 1985 (in Polish).
33) Sakakima, H., Yanagiuchi, Y., Satomi, M.,Senno, H. and Horota, E., ref. 11, p. 941.
34) Nathasingh, D.M., J. Appl. Phys., 55, 1793 (1984).
35) Mohri, K. IEEE Trans. Magn., MAG-20, 942 (1984).
36) Mohri, K., ref. 2, p. 1687.

37) Harada, K., Sasada, I., Kawajiri, T. and Inoue, M., IEEE Trans. Magn., MAG-18, 1767 (1982).

38) Sasada, I., Inoue, M., Hiroike, A. and Harada, K., IEEE Trans.Magn. MAG-19, 2148 (1983).

39) Sasada, I., Hiroike, A. and Harada, K., IEEE Trans. Magn., MAG-20, 951 (1984).

40) Kwiatkowski, W., Konopka, J., Kozak, M. and Misiuk, E., Research Report PR-3.19.2 3, Warszawa, Nov., 1985 (in Polish).

41) Mohri, K., Takeuchi, S. and Fujimoto, T., IEEE Trans. Magn., MAG-17, 3370 (1981).

42) Mohri, K. and Takeuchi, S., J. Appl. Phys., 43, 8386 (1982).

43) Masumoto, T., ref. 11, vol. I, p.5.

44) Spano, M.L., Hathaway, K.B. and Savage, H.T., J. Appl. Phys., 53, 2667 (1982).

45) Arai, K.I. and Tsuya, N., J. Appl. Phys., 49, 1718 (1978).

46) Mendelsohn, L.I., Nesbitt, E.A. and Bretts, G.R., IEEE Trans. Magn., MAG-12, 924 (1976).

47) Filipensky, J., Svestka, Z., Tischer, Z. and Fiola, I., J. Magn. Magn. Mat., 41 , 437 (1984).

48) Dismukes, J.P. and Sellers, G.J., Proc. 3rd Int. Conf. Rapidly Quenched Metals, Brighton, vol.II,p.205, 1978.

49) Borek, L., Zeitschirft ELEKTRONIK, nr 4, 1982.

50) Hilzinger, H.R., ref.2, p.1695.

51) Gonser, U., ref.2, vol. I, p.XLVII.

Table 1

Material	B_s(T)	B_r(T)	ρ(μΩm)	d(μm)	FOM x 10^{-3}
2605 CO[1)]	1.80	1.60	1.30	25	24.0
2605 S-2[1)]	1.56	1.3	1.30	25	17.0
3%Si-Fe[2)]	2.0	1.6	0.50	25	10.4
50%Ni-Fe[2)]	1.6	1.5	0.45	25.	6.9

[1)] data taken from "METGLAS Electromagnetic Alloys", Allied Co.

[2)] after ref. [22)].

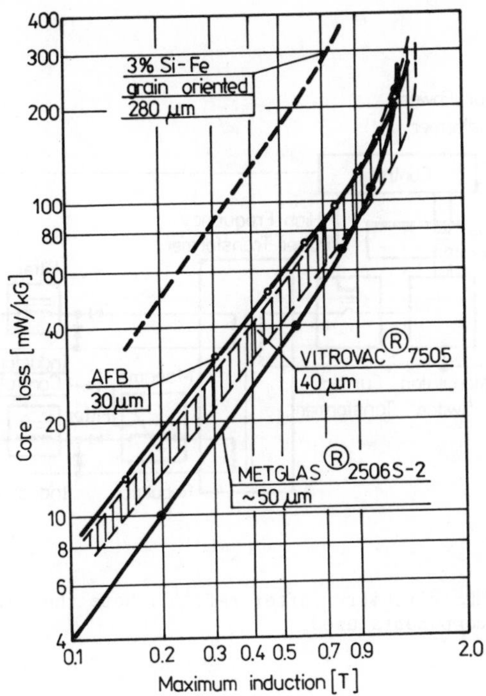

Fig.1. The 50 Hz core loss characteristics of the Fe-based metallic glasses manufactured by different producers, and for comparison, also the one of 3% Si-grain oriented steel.

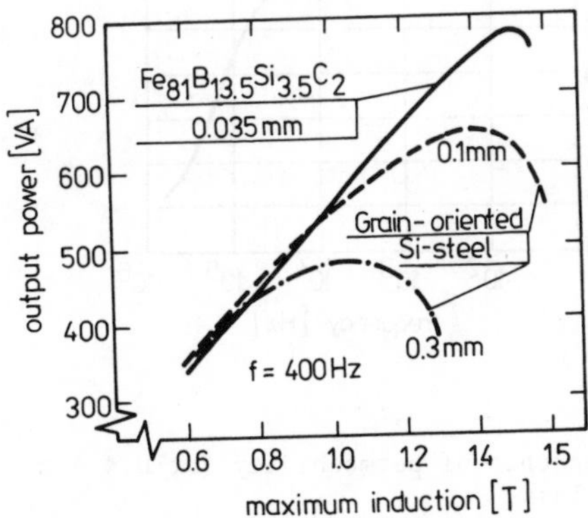

Fig.2. Power output of a 400 Hz transformer with torroidal cores of Si-steel and Metglas 2605 SC (after ref.[21]).

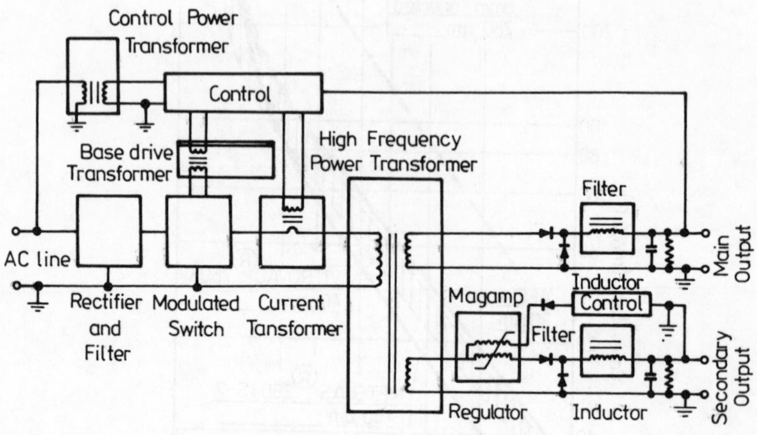

Fig.3. Typical SMPS circuitry (after ref. [1]). Note the variety of magnetic components used.

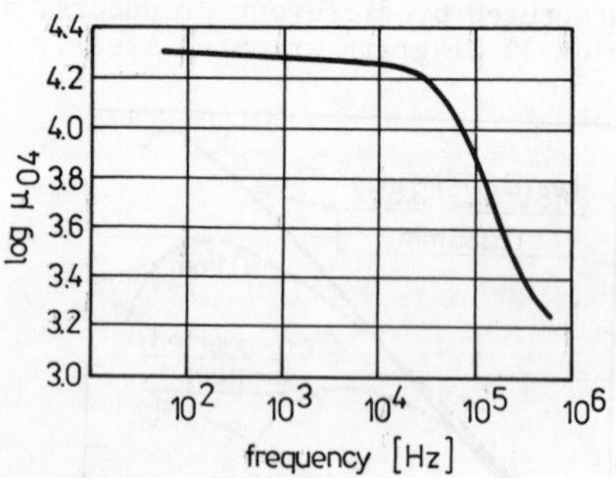

Fig.4. Frequency dependence of permeability (at 0.4 A/m) of the AMM-metallic glass.

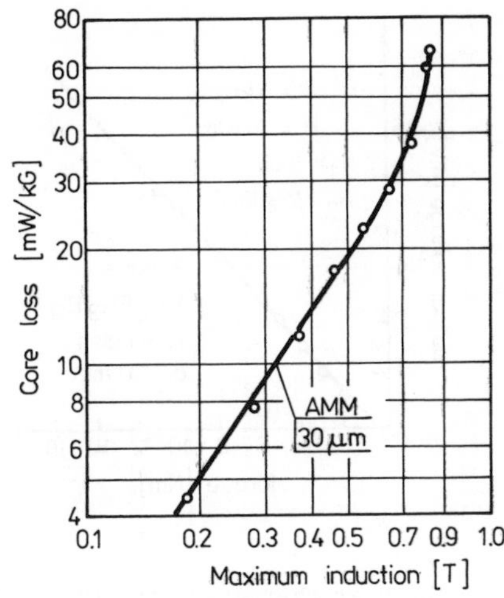

Fig.5. The 400 Hz core loss characteristic of the AMM-metallic glass.

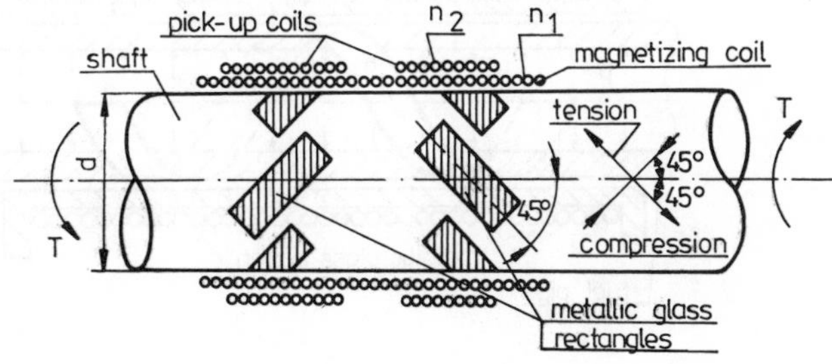

Fig.6. A schematic diagram illustrating the idea of a torque transducer (after ref.[39]).

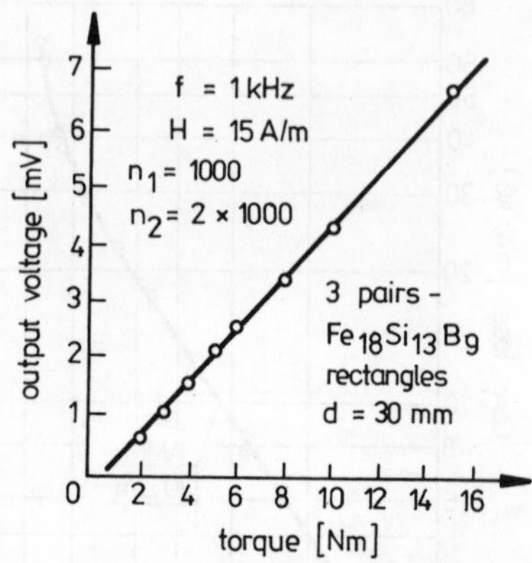

Fig.7. Torque transducer characteristic (by courtesy of Mr. T. Walecki from the Institute of Physics, P.A.Sci.).

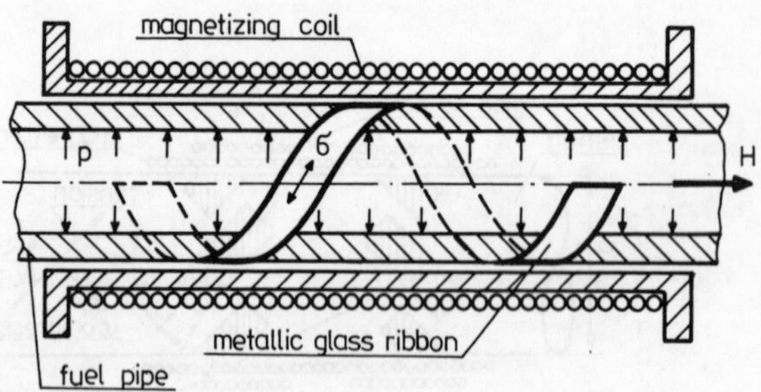

Fig.8. Schematic diagram of the angular velocity transducer of a Diesel engine (after ref [40]).

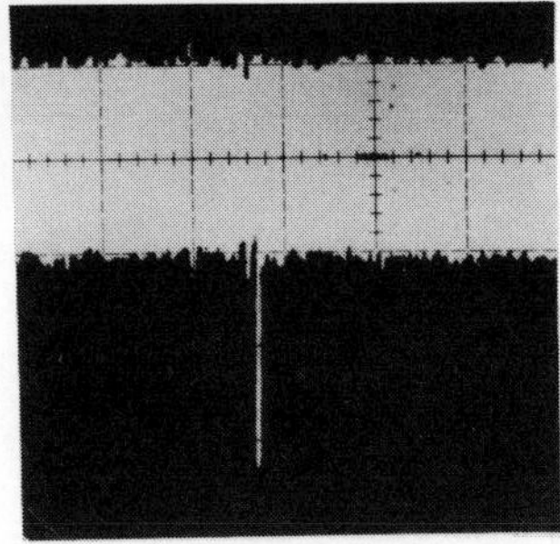

Fig.9. Typical output pulse obtained in the laboratory model of the transducer shown in Fig.8; vertical cal. 2 mV/div. (by courtesy of Dr. J. Konopa from the Institute of Theoretical Electrotechnique and Electrical Measurements, Warsaw University of Technology).

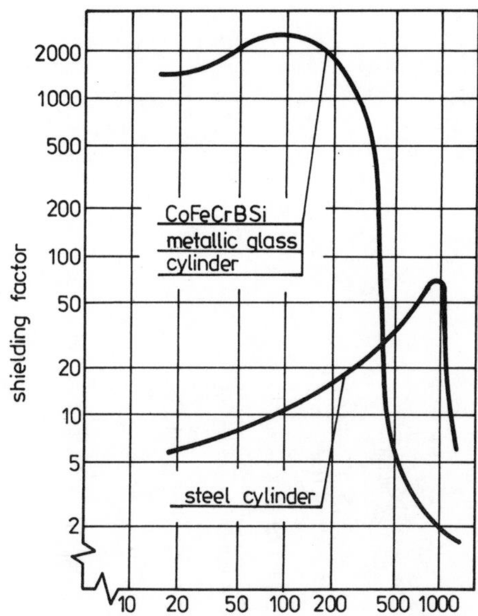

Fig.10. The dependence of the shelding factor of model screens made using the CoFeCrBSi-metallic glass and steel (after ref. [47]).